Breeding Bio Insecurity

How U.S. Biodefense is Exporting Fear, Globalizing Risk, and Making Us all Less Secure

Lynn C. Klotz
Edward J. Sylvester

16
EasyRead Large

RHYW

Copyright Page from the Original Book

LYNN C. KLOTZ is senior science fellow with the Center for Arms Control and Non-Proliferation. He has published over forty research papers and review articles in leading scientific journals and numerous articles on biotechnology business and biological weapons arms control.

EDWARD J. SYLVESTER is a science journalist and author of four books on research and medicine in the age of the "new biology." He teaches science and medical writing in the Walter Cronkite School of Journalism and Mass Communication at Arizona State University.

The University of Chicago Press, Chicago 60637
The University of Chicago Press, Ltd., London

Printed in the United States of America

18 17 16 15 14 13 12 11 10 09   1 2 3 4 5

ISBN-13: 978-0-226-44405-5 (cloth)
ISBN-10: 0-226-44405-8 (cloth)

Klotz, Lynn C.
Breeding bio insecurity : how U.S. biodefense is exporting fear, globalizing risk, and making us all less secure / Lynn C. Klotz and Edward J. Sylvester.
p. cm.
Includes index.
Summary: This book argues that the conditions of research in bioweapons and biosecurity pose a greater risk to health and security of Americans than do bioterrorist attacks, but that this risk can be countered and defeated with greater efforts against infectious diseases and greater international oversight and transparency.
ISBN-13: 978-0-226-44405-5 (cloth : alk. paper)
ISBN-10: 0-226-44405-8 (cloth : alk. paper) 1. Biological warfare. 2. Biological weapons—Research—United States. 3. United States—Military policy—21st century. 4. Communicable diseases—Prevention. 5. Public health—International cooperation. I. Sylvester, Edward J. II. Title.
UG447.8.K57 2009
358'.38—dc22 2009013803

♾ The paper used in this publication meets the minimum requirements of the American National Standard for Information Sciences—Permanence of Paper for Printed Library Materials, ANSI Z39.48-1992.

Set in 16 pt. Verdana

TABLE OF CONTENTS

Acknowledgments

We would like to acknowledge the many experts who have read and commented on the manuscript. Special thanks to Martin Furmanski, who commented on several chapters and generously supplied unpublished material from his own research. And special thanks to Ambassador James Leonard, whose careful reading and rereading of chapter 3 was invaluable to assure its historical accuracy. Thanks to Jens Kuhn, Mark Wheelis, Marie Chevrier, John Gilbert, and the anonymous reviewers of the manuscript, all of whom caught errors and whose suggestions certainly improved the book. And finally, thanks to Chandré Gould for sharing her insights on Project Coast. Of course, we are responsible for any remaining inaccuracies. Lynn Klotz would also like to thank Ruth Melchior Fleck for her support during the writing of the book.

Lynn Klotz would like to dedicate this book to his family: Cassandra, Kevin, Zewde, and Seif.

Ed Sylvester offers his dedication to wife Ginny and to their son Daniel, daughter, Kate Broadman and her husband Anthony, and their granddaughter Nica.

CHAPTER ONE

Dangerous Crossing

In the autumn of 1347, a Mongol army attacked the Italian trading outpost at Caffa in the Crimea. While laying siege to the city, the invaders began to die in large numbers from an unknown but incredibly virulent disease. That ended the attack, but legend has it that before the Mongols departed, they catapulted the bodies of victims into the city in what would be, if true, one of the world's deadliest—and therefore most successful—uses of biological warfare.[1]

When the siege survivors arrived in Messina, Sicily, legend ends and history begins. Those who greeted them were horrified to find the sailors dead or dying at their oars from a disease that hideously blackened their skin and caused egg-sized bleeding pustules—or buboes—to form in their armpits and groins. Soon the onlookers became victims, then those who cared for the onlookers, then those who buried the caregivers. So began Europe's bubonic plague that would kill half the populations of its largest metropolises—Paris, Florence, and Venice. Striking equally across social classes, it would change dynasties, reshape social life and trade, and alter history in ways we can never know.

Within a few years, the plague killed a third of the continent's population and a similar percentage in a swath ranging from China to India, through the Byzantine and Ottoman empires, all the way to the Middle Eastern Levant, each offering yet another port of entry to Europe's crowded cities. It devastated the entire known world. And if it marked the first European encounter with biowarfare, it certainly would not be the last.

At each successive step, the perpetrators of biological warfare grew more knowing. When the British outpost of Fort Pitt was threatened by Delaware tribesmen in 1763 during the French and Indian War, the defenders distributed blankets and handkerchiefs that had been used by smallpox patients among the natives—a telling reversal of roles from the Caffa siege.[2] That much was recorded by one of the fort commanders. The entire American Indian population ultimately was devastated by smallpox, though it had already been destroying native populations for two centuries by the time it was used at Fort Pitt. Three decades later, Dr. Edward Jenner created the vaccine that brought smallpox to bay in the developed world.

From then on, as the science of microbiology advanced, biowarfare crept forward in its shadow. During World War I, the Germans attempted to kill Allied mules and horses by infecting them with laboratory strains of anthrax and glanders, both bacterial diseases, in order to disrupt military supply lines.[3] It was mostly unsuccessful and had no effect on the Allied war effort. In World War II, the Japanese killed tens of thousands of defenseless Chinese in occupied Manchuria over a period of years in perhaps hundreds of "experiments" with plague, cholera, and other deadly bacteria sprayed from airplanes, put into food and water wells, and injected directly into victims. Given the Chinese victims truly were defenseless, tens of thousands of deaths is not surprising or even evidence of the mass killing power of bioweapons.

But in the twenty-first century, all prior efforts could be dwarfed by creations of "the new biology": the marriage of molecular biology, which has brought profound understanding of the molecules of life, and biotechnology, its practical complement. The discoveries and creations announced daily that lead us toward a world of miracle cures and preventatives may bring, in lockstep, arrays of bioweapons powerful enough to quite literally hijack our minds and bodies. A most important truth of historical biowarfare is not the devastation it accomplished, but its limitations.

When physicists exploded the first nuclear bomb at Alamogordo in 1945, they witnessed a horror no human had ever seen. In sharp contrast, earlier bioweapons never were a match for natural diseases, which can approach the horror of even nuclear weapons. The 1918 influenza pandemic killed forty million people worldwide, far more than died in World War I.[4] When smallpox was eradicated from Earth thirty years ago, a scourge ended that over the centuries had killed and maimed hundreds of millions of people and destroyed over half of populations lacking resistance.

Moreover, though sometimes dramatic and immediately successful, biowarfare attacks of the past simply struck an enemy with diseases that either were well known and sometimes endemic, as in the Japanese experiments, or that were already on the march. The plague introduced with such lethal effect at Messina already hovered at Europe's borders and soon stormed in from everywhere. Smallpox carried by Europeans had been killing native populations unintentionally before the British struck with it at Fort Pitt. Historic biowarfare was hit or miss.

Now imagine the monumental history of natural biological disaster repeating in warfare, the agents delivered not clumsily but with the full force of natural pandemics, their delivery and even their lethality radically enhanced by science. That kind of attack would strike like a hurricane that has built up power over the ocean, sweeping in with predictable malevolence but unimaginable force. That energy-rich ocean—a wilderness as full of promise and danger as any ever explored—is the new biology. It is impossible to talk about the dangers inherent in biowarfare without looking toward the new knowledge of microbes and human defenses gained over the past half century, and it is equally impossible to talk about the wonders of modern genetics and drug discovery without attending to their dark potentials.

If that weren't enough, it is equally impossible to discuss the dangers of human-manipulated organisms without conjuring the complex natural world of evolving microbes such as avian flu virus, which threaten evermore havoc as we invade their natural reservoirs and unwittingly spread them throughout the globe.

How to engage these problems in the ways needed to reach an attainable level of security against all these biological threats is the theme of *Breeding Bio Insecurity.* Biosecurity is as complex as all this implies. There are no simple solutions. There *are* realistic strategies that can reduce the threats. And most important, in many cases our government strategies take us directly away from them toward the dangerous future just portrayed.

Haste

In the years following 9/11, the United States has poured out billions of dollars for massive expansion of high-biosecurity labs and has encouraged universities and private sources to build them. The irresistible lure is millions in research funding to develop countermeasures for weapons feared to be on the horizon, the goal to build walls of protection against bioterrorism or the full-scale biowarfare that might be waged by rogue nations. And though reacting to those frightening prospects, at the same time the government has clamped a lid of security on such work, sometimes accompanied by draconian punishments for violators.

Intuitively, all that may seem an unfortunate but necessary response in the aftermath of terror. Americans have good reason to be apprehensive about bioterror and other forms of "asymmetrical warfare." First came the shock of 9/11, witnessed in real time by perhaps a billion people around the world. On its heels just weeks later, the anthrax attacks left five people dead, more permanently injured, and all of us with foreboding that this was just a whisper of what true biowarfare might bring at any moment. The government struck back with unprecedented intensity and haste to fix blame, seek revenge, and cobble together a monumental biodefense program. And that is the heart of the problem. Intuition in the heat of disaster is not a good shaper of policy for the most powerful nation in the world dealing with one of the most complex issues of our time—nor would it be for any nation at any time.

On balance, we may be less biosecure than before the 2001 attacks. Our bloated, largely secret biodefense program increases the risk of accidents and theft by terrorists, and its lack of transparency may be inadvertently fueling an international arms race in bioweapons. The danger of starting a pandemic from the escape of deadly viruses from a lab may far exceed the medical benefits that might accrue. Much-needed oversight of dangerous activities is insufficient or nonexistent.

And as fearful as science-enhanced biological warfare would be, the likelihood of massive attacks by rogue nations or terrorists is low, now and for years to come, and does not warrant the enormous concentration of resources to protect ourselves at any cost that appears to be the goal of U.S. biodefense policy. More seriously, that policy makes the onset of such biowarfare more likely than it otherwise would be. Consider that anomaly, for you will see the evidence of it mount up chapter by chapter—evidence we believe far outweighs the countervailing claims of other biosecurity experts we will discuss.

Buying a Deadlier Future?

Perhaps a quarter of the nearly $50 billion in the U.S. biodefense program goes to research and develop bioweapons countermeasures like antibiotics, antivirals, antidotes, and vaccines, in a rush to protect us from bioterrorism. Testing them clearly requires ready availability of the bioweapons agents themselves. On the other hand, the very presence of agents like weaponized anthrax and plague in hundreds of high-biosecurity labs is, just as clearly, an open invitation to theft.

To garner shares of the countermeasure funding, a swelling cadre of academic scientists is beating a path to research Earth's deadliest microbes and toxins to see how they kill in order to develop countermeasures. The government's goal is defense. Because such work requires highly protective environments, the would-be developers have rushed to build or refocus hundreds of laboratories of a type known as Biosafety Level 3. Biological safety is rated on a scale in which the top level, 4, demands the most extreme precautions, required in order to work on the deadliest microbes, whose escape could have the direst consequences. There are now over one thousand BSL-*3* labs registered to work on "select agents,"[5] maybe several hundred of them working with bioweapons agents. The United States plans for even more BSL-*4* labs; there are already fifteen operational or nearly so, up from only four a few years ago. The tales of errors that have already occurred with deadly microbes in BSL-*3* labs and other settings are hair-raising. Whatever inherent dangers they pose, how those hazards increase with proliferation is easy to demonstrate.

If the risk of a deadly microbe's being lost, stolen, or somehow escaping from one BSL-*3* lab at the historic bioweapons research facility of Fort Detrick, Maryland, was *x,* what is the danger of that mishap when there are more than a thousand such laboratories? An immediate answer would be at least a thousand times *x,* but it is actually likely to be much greater, because most of the new labs will be staffed by personnel relatively untrained in handling deadly microbes.

Richard Ebright, a professor in the prestigious Waksman Institute of Microbiology at Rutgers University, says that security measures are not nearly tight enough to protect against theft of organisms. Ebright noted, "If al-Qaeda wished to carry out a bioweapons attack in the U.S., their simplest means of acquiring access to the materials and the knowledge would be to send individuals to train within programs involved in biodefense research." Regarding background checks, he points out that "[9/11 mastermind] Mohammed Atta would have passed those tests without difficulty."[6]

Wages of Fear

These federal biodefense activities in the name of security in fact are putting all of us at ever-greater risk. We are at greater risk in our neighborhoods, and we are at greater risk internationally of being perceived as aggressors.

James Leonard, Richard Spertzel, and Milton Leitenberg, three of the country's most respected experts on the state of biosecurity in the world today, warn: "The rapidity of elaboration of American biodefense programs, their ambition and administrative aggressiveness, and the degree to which they push against the prohibitions" of the long-standing international agreement against developing biological weapons "are startling."[7]

The three have established and impressive U.S. biosecurity credentials, two for work with the government. James Leonard was the ambassador who headed the U.S. delegation in forging the international agreement, the Biological Weapons Convention (BWC), during the Nixon administration, and Dr. Richard Spertzel was former deputy director of the famed U.S. Army Medical Institute on Infectious Diseases—a unit whose vision is to be America's leading research laboratory in "providing cutting-edge medical research for the warfighter against biological threats."[8] The third is Professor Milton Leitenberg, senior research scholar at the University of Maryland's Center for International and Security Studies. He notes that if the United States observed any other country engaged in or simply planning the kinds of biodefense effort that $50 billion is buying, we would judge that country to be violating the BWC, the prohibition against bioweapons development that has stood for more than thirty years.[9]

President Richard M. Nixon, renouncing the development of offensive biological weapons in a move that led to adoption of the BWC, said that mankind "already carries in its hands too many seeds of its own destruction."[10] He felt that biological warfare had "massive, unpredictable, and potentially uncontrollable consequences" and biological weapons were "repugnant to the conscience of mankind."

However repugnant, we are courting the disease itself in the name of prevention. The risk of an international biological arms race launched by the obsessive secrecy that is the U.S. norm can only compound whatever threat is already out there. How better to launch such an arms race than to give every appearance of starting one yourself while keeping others in the dark?

We are charging down this dangerous path out of contradictory motivations that can only be self-defeating. The government is paranoid over potential bioterrorism while being permissive toward those whose work might well fuel it. Paranoia is a keystone in our government's political policy of instilling fear to maintain an image of being strong on terror. This policy has created an illogical fear of a massive terrorist attack with biological or chemical weapons. At the same time, billions in biodefense funding have been made available—with billions more planned—with little thought to consequences, to entice scientists to develop countermeasures to protect us from a massive biological weapons attack that may never come. With little control or even oversight over how this money will be spent, we are creating a permissive atmosphere for unthinking scientists to dream up whatever research with dangerous pathogens might help them get the money. And all that makes public fear warranted, though hardly in the ways intended.

Steering Clear of Danger

If so much of the U.S. effort is wrong-headed, how *do* we achieve biosecurity in this century of the new biology? That will mean security against classical biological weapons and hostile exploitation of the new biology; security from emerging and reemerging pandemic flu, antibiotic-resistant infections, and ever-present infectious diseases; and, hardest to articulate, security against yet-undiscovered threats.

A number of strategies now in play or on the horizon would increase biosecurity, most obviously research and development (R&D) on biological weapons agents and countermeasures and, as we will argue, even greater efforts that should be waged against natural infectious diseases. But equally critical goals include preventing the hostile exploitation of biology, from ongoing international efforts toward treaties and agreements to grassroots activities such as developing ethical awareness in scientists. Finally, in order to reduce the risk of an arms race in bioweapons, efforts to bring international oversight and transparency to biodefense activities and dangerous research are crucial.

Oversight and transparency may have the kind of "policy-wonk" ring that makes your eyes glaze over. But apply them to your street and they ring with urgency. Suppose you learn that the building going up down the block is a laboratory to research the deadliest infectious diseases in the world. Questions jump to mind: Precisely what will go on in there? That's transparency. How can I be sure no experiments now or in the future risk my neighborhood's safety? That requires oversight, proactive rather than after work is underway, and continuing indefinitely.

Residents of Boston's Roxbury section learned those lessons quickly and well when Boston University proposed building a BSL-*2,* BSL-*3 and* BSL-*4* lab complex right in the middle of their neighborhood. The ongoing battle for approval has pitted the city's elected officials, congressional delegation, and the university against the residents, who enlisted powerful allies from a quarter that surprised many casual observers. Petitioning against the project were 147 leading scientists, including two Nobel laureates and several leaders in the U.S. biosecurity community. They argued that placing the lab in the center of densely populated Boston neighborhoods was a dangerous idea.

Good questions for the lab down the street are good for nations to ask of others a continent away.

Oversight and transparency assure that biodefense work is in fact defensive and that it is conducted safely. Research with dangerous endemic and exotic infectious disease agents cries out for the same oversight to assure that it is being done safely.

We argue that true biosecurity can emerge only from policies that protect and advance infectious-disease public health, not from those aimed simply at biowarfare. Except for well-funded research on a few diseases, such as AIDS and potential pandemic flu, new money is desperately needed to combat major emerging and reemerging public health threats, such as tuberculosis and drug-resistant staph infections, as well as garden-variety flu. Furthermore, many genuine public health threats, such as AIDS and antibiotic-resistant staph infections, usually require more modest biosafety containment—the BSL-*2* found in countless university labs—because the diseases already are endemic in the United States and are not particularly infectious in a lab setting.

Most research on deadly pathogens like the Ebola and Marburg viruses can be defanged by working with strains made noninfectious using genetic engineering or working with single molecules from the pathogens. Generally, such research can be conducted at BSL-*2* or even BSL-*1* labs. Only when vaccines or other countermeasures must be tested would the live infectious agent and high biocontainment be required.

Amazingly, the 1918 influenza virus that killed forty million people does not require complete, highest security BSL-*4*—although it certainly should, in our opinion. It requires a containment "hybrid" between BSL-*3* and BSL-*4*.

Finally, anthrax is not normally fatal to humans, and there are antibiotics and a vaccine against it, so it usually requires only BSL-*2*—*except for aerosol experiments on weaponized deadly strains and experiments that involve infecting large animals.* And those exceptions offer another major reason so many biosecurity experts are alarmed. Boston's and other BSL-*3* and BSL-*4* labs might well be researching biodefenses against Ebola, Marburg, and weaponized anthrax, but it certainly will not look that way to enemies, real or potential. Such work looks like a cover for offensive bioweapons development, and because of the secrecy increasingly shrouding bioweapons research, there will be no way for us to prove otherwise.

None of this implies that we do not face new, genuine threats. We do, and we must prepare for them. Just as science's harnessing the atom allowed us to *un* harness its unprecedented destructive power, the lifeenhancing powers of the new biology lead directly to the dangers of hostile exploitation. While the United States may be radically overestimating the current risks from biological and chemical weapons, developing them will become increasingly easy and the weapons increasingly sophisticated. That is the dark side, where the understanding of life processes and the ability to control and alter them, sought to cure disease and extend our lives, instead may be directed toward hostile exploitation.

Let us instead put bioweapons threats in perspective, because seeing them in perspective offers a chance to understand why this massive government response is unwarranted:

- In the anthrax letter attacks, five people died, yet as its potency is measured, there was enough anthrax in those letters to kill hundreds, maybe thousands. Hundreds more did not die because that potency is measured under "ideal" conditions of a laboratory, which cannot even be approached outside its controlled environment.
- Fatality estimates in the hundreds of thousands following a massive attack that might unfold soon are not based on real evidence or data. There never has been a single attack that resulted in

massive casualties, despite Japanese efforts to do so in China.

- The number of offensive biological weapons programs in the world has decreased over the last several years.
- Launching a high-fatality attack is certainly well above the technical capabilities of terrorists. To safely produce and weaponize biological agents requires a safe lab facility and considerable technical expertise, not something that terrorists—or even most developing nations, for that matter—can muster on their own.

However, if we simply consider the lethality and morbidity or contagiousness of the agents themselves, we come to a point of universal agreement. Anthrax and the other major bioweapons agents complete a cast of microscopic horrors to populate the worst nightmares, and later it will be clear why. Terrible as they might be, the natural outbreaks and epidemics of these agents only kill some *percentage* of their victims—nothing like, say, a hydrogen bomb. But the outcome of warfare conducted with microbial agents recrafted with the powerful new tools of biology could be another story.

Under that scenario, many scientists fear fatality rates that rocket off the charts. Biological warfare deaths and permanent casualties may grow to the same unthinkable magnitude as those of nuclear war. They could truly become "weapons of mass death"—not of mass destruction, because microbial and chemical agents do not destroy buildings or, if carefully designed, leave land uninhabitable. Just as in a nuclear attack, some would survive, but they likely would find themselves in the world forecast by Soviet leader Nikita Khrushchev, who reportedly said that following a nuclear exchange, "the living would envy the dead."[11]

Are we Americans working to defend ourselves against the day when bioweapons become a big threat? Or are we secretly and unwittingly inching our way toward creating our own offensive bioweapons? No bright line separates the two outcomes. There is only shadow and blur, the very techniques used to induce fear and suspense in horror movies. As more and more research creeps into secrecy in the name of national security, the chances dim of ever assuaging the fears of foes or even friends. Eventually, even scientists might not know that they were working to create offensive bioweapons—not if the past is prologue.

Russian Roulette

In addition to the risk of spurring an international biological arms race that so much of our work creates, much of the research being undertaken is hazardous in itself. Most dangerous experiments originate in the pursuit of biodefense or from the increasing worry that a natural pandemic will kill hundreds of millions worldwide. Witness scientists' re-creation of the lethal, highly contagious 1918 influenza virus and its continued study in labs, a singularly controversial achievement whose ramifications must be explored. Still others began with investigations thought to be harmless that went wrong, as in a case famous among researchers in which Australian scientists' efforts to use a relatively benign mousepox virus to make mice healthily sterile ended up killing them, and suggested a similarly lethal pathway for the human pathogens cowpox and smallpox. As in that accidental discovery, some were followed up by work deliberately researching newfound hazards in the name of building protection against them. However, escape from the lab of some of these dangerous pathogens may unleash the dreaded pandemic that the microbes are being bred to prevent, and while many in government attempt to downplay the risks of accidental escape or lab worker infection, they are flying in the face of recent history that shows how very easily, and often inexplicably, it all happens.

Errata

The microbes studied in this country's thousand-plus BSL-*3* labs can be every bit as deadly as those housed in BSL-*4*. Under the U.S. classification system, the diseases they cause are simply treatable. Treatable includes bubonic plague and anthrax in their early stages, tularemia, and a host of others you would not want for neighbors—but which you may indeed have, if not already, soon.

For example, in 2005 three research mice infected with deadly bubonic plague bacteria disappeared from a BSL-*3* lab in Newark, New Jersey. Nobody knows what became of them, whether the loss represented an accounting error or they were cannibalized by other mice, or whether in fact they escaped into tenements of the surrounding low-income community. Officials at the lab assured the public that the whole issue was in fact an accounting error. It would not be the last time for such assurances at the same lab.

More recently, at Texas A&M University a lab worker was accidentally infected with a microbe being studied under supposedly secure conditions, and she spent weeks home sick before being appropriately diagnosed. In an earlier incident only reported after the first was uncovered, workers had been exposed to a bioweapons agent, though they did not become ill. Most significantly, the late reporting of both was in violation of federal law, providing strong evidence that not reporting even dangerous accidents may be the rule rather than the exception. No one knows how many lab accidents go unreported, but many of those that are known are hair-raising indeed.

Trust—Don't Verify

The world had a singular opportunity to see complex, hidden activities of science and government laid bare with the end of apartheid in South Africa. The nation's Truth and Reconciliation Commission provided participant and eyewitness testimony to experiments conducted under situations remarkably similar to those in the United States today. The story is all the more chilling because the former government demonstrably had aimed to create offensive biological weapons—for use against its own black citizens.

Many scientists were recruited for defensive biological weapons research, which was clearly allowed under the world's Biological Weapons Convention. At least, many insisted they truly believed the work to be purely defensive. They were easily recruited for that patriotic goal. Though at peace, white South Africans at the close of apartheid perceived security threats from a Communist takeover, from invasion by neighboring countries, from their own black countrymen. And in addition to their patriotic feelings, many scientists were enticed by a passion for science and by their own personal ambitions.

President Ronald Reagan famously said of the need to monitor Soviet arms development despite the Russians' assurances of peaceful intent, "Trust, but verify." How could South Africa's scientists have verified what they were really producing at their government's behest when the work was shrouded in secrecy? How could any scientist or any one of us?

Secrecy and paranoia have driven the United States to abandon the moral high ground that we held for decades in controlling biological and chemical weapons, instead increasing the likelihood of an arms race in biological weapons through aggressive and hidden biodefense activities. We are driving experienced bioweapons researchers away from work on countermeasures for bioweapons agents through draconian rules and overzealous enforcement while attracting inexperienced young researchers and encouraging unsafe and dangerous research projects by supplying billions of dollars in funding for them. Finally, we are losing focus on the most critical public health areas, where new funding is needed for emerging and reemerging infectious diseases as well as those that harm and kill with predictable regularity.

Déjà Vu: The Right Side of the Looking Glass

The goal of this book is to show how we can avoid the deadly mistakes of the past as we delve into the many sound ways to protect ourselves against hostile exploitation and dangerous activities of the new biology, a view that is shared by many biosecurity experts. But it will take concerted attention and effort from grassroots groups to scientists to policy makers to achieve. Of the highest importance are transparency, public oversight, and international cooperation. Unilateral action and secrecy are the ingredients of paranoia. Paranoia leads to arms races. Oversight can help prevent dangerous activities before they start. And we are all in this together, so international cooperation is critical. In this age of rapid global travel by millions of people of all nationalities, an attack against one country is truly an attack against everyone.

Fortunately, the means to international cooperation have been developing over the past century, beginning with the Geneva Protocol in 1925, a reaction to the horror of the gas attacks of World War I, when poison gas killed 91,000 people and left 1.2 million casualties, many maimed for life. With foresight, the protocol also banned the use of biological weapons. That was only the beginning, but it was an opening in which two major powers, the United States and Japan, did not take part. The United States finally ratified the protocol fifty years later, in 1975.

The protocol banned only the use of chemical and biological weapons—not their development, manufacture, or stockpiling. But if you build stockpiles, it is too tempting to use them. The last important steps to ban development, production, and stockpiling were left to the BWC, which became law in 1975, and the Chemical Weapons Convention (CWC), which became law only in 1995. There are loopholes, however. Mounting a defense against a biological weapons attack is an allowed "protective" purpose, but activities that enable that defense are only vaguely defined. Unfortunately, our government is using this vagueness to justify activities that some believe are pushing up against the prohibitions of the BWC. We believe that the United States should take the moral high ground and demand of itself and others the strictest interpretation—that we should be "strict constructionists" of the Biological Weapons Convention.

The BWC also introduced a new concept that must be considered to keep the world safe from these dark uses of new biology. It bans developing the weapons not only for warfare but for other "hostile purposes." Why is the last step needed? Because even without resorting to warfare, diseases or biotoxins could be used to assassinate enemies, to disrupt a country's tourism, or to destroy its food-export market. Certainly this would amount to state terrorism, but there is equally fearful potential for "commercial terrorism," in which a company would seek to undermine a rival with pathogens.

The sister Chemical Weapons Convention bans use in warfare of both lethal and nonlethal chemical weapons, which would include the scariest weapons of all: mind-control agents. However, riot control agents, which conceivably could be replaced with mind-control agents, are allowed for domestic purposes. Could they be used to pacify citizens of an occupied territory? Perhaps. The chemical treaty is vague on that point, and other treaties actually require occupiers to enforce the law.

In the end, we believe that only ethically aware life scientists can save us; they are the first line of defense. They will be the first to witness dangerous or hostile research and development activities, and they will be the ones recruited to carry out those activities. Scientists stand alone. They are not protected by labor unions, they are more likely to be sworn to secrecy agreements than others, and they have little whistleblower protection. It is of paramount importance to provide ethical training and protections for them.

Next, informed citizens must distinguish between policy proposals that build protection against genuine public health threats and those that use fear and alarmist tactics to lead away from biosecurity while claiming to protect.

Scientists, citizens, policymakers. Each of us is essential to protecting all of us. No one group or singular effort will accomplish this monumental but essential task, nor will there be a final victory. Just as over the decades the citizens of the world have maintained a constant vigilance against nuclear threats, we now need to awaken to the dark sides of twenty-firstcentury biology as well as its dazzling promises.

Think globally, act locally. So exhorted the scientist/environmentalist René Dubos. However, like every idea that occupies the border between science and policy, this vision of citizen and scientist activism roils with countercurrents. The government's fear of global terrorists and other enemies leads to danger in our backyards. And the eruption of high-security biosafety labs that threatens our neighborhoods breeds fear in our allies, let alone our enemies, worldwide. Every unwarranted move toward secrecy brings us incrementally closer to an international biological arms race whose menace not only begins at home, it ends at home. In a world of supersonic-weapon delivery systems and near-supersonic travel, we all live in one village. In the twenty-first century, we also must turn Dubos's message around: By thinking locally, we will protect the globe.

All this is taking place against a backdrop of the accelerating pace of new capabilities in biology, which eventually will allow scientists of routine skills with hostile intentions to alter pathogens—and even humans.

Far into the future we will be living in a world of choices based on the dual uses of biological discovery. We are urging attention to what prominent biosecurity experts believe are wrong uses because they dominate virtually every aspect of U.S. policy. And because every dark prospect we foreclose opens the door to its opposite number, a development that might truly benefit humans and the rest of the living world.

CHAPTER TWO

A Future Bright and Dark

Imagine you put on earphones while you watch a multimedia slide presentation by a scientist who is expert in biosecurity issues. As each newly discovered wonder of the human genetic blueprint emerges in color on the screen, he or she is going to predict an "exciting" outcome—and that word has never had so ambiguous a ring. If you turn the headset dial to "green," with each slide you will hear of a benefit to health, happiness, or another hallmark of human progress. But if you set the dial to "red," the same slides will bring the sort of dire warnings we've just been discussing: of a new weapon that could threaten you and your family, of experiments gone out of control, or even that worst of outcomes, mind-control molecules that would change who you are by someone else's decision. Red or green?

Never before have two opposing directions been so clearly implied in every scientific discovery. We must choose whether to exploit the beneficial or hostile path into the future. Nothing less than the course of civilization is at stake.

Discoveries in molecular biology are changing our understanding of human complexity. Unraveling these newfound complexities requires emerging disciplines such as systems biology, which represents the integration of computer science, mathematics, physics, chemistry, and other disciplines into one broad-based specialty of biology. Synthetic biology unites life scientists with engineers from a variety of those disciplines. Other new subdisciplines represent the opposite, the differentiation of existing specialties in order to intensify a research focus. For example, the rapid pace of drug discovery led to many genomic subspecialties referred to in shorthand slang as "omics." There is proteomics, focusing specifically on the functions of proteins; metabolomics, to study how proteins work together in metabolism to produce the molecules the body needs; toxicogenomics, to pinpoint causes of chemical toxicity and to circumvent such problems in new drugs.

Finally, pharmacogenomics attempts to answer some of the most critical and frustrating questions in all of medicine: Why does a potential wonder drug for 99 percent of the population have devastating effects on 1 percent, limiting or eliminating its use if we cannot identify the ill-fated patients in advance? Why do some drugs completely cure 50 percent of subjects in a clinical trial and do nothing for the rest? Often the answer lies in our slightly varying genes, which give rise to different reactions to drugs.

New tools helped forge the new specialties, and some of these were only recently the stuff of science fiction:

- Microarrays the size of a postage stamp can analyze your entire genome in days, until now nearly impossible in any time frame.
- Imaging devices visualize single molecules inside cells so their activities and chemical fates can be studied.

And on the horizon:

- Ultra-rapid DNA sequencing methods will enable all of us to carry around our complete DNA sequences, so they will be available for doctors to design individually tailored therapies.
- New tools and methods will offer near instantaneous disease diagnosis at the bedside.

Sometime in the future we will use machines and computers on the submicroscopic scale of molecules, which is what they will be.

All this sounds positive and upbeat, but the future will only play out this way if we use the knowledge and tools for beneficial purposes rather than hostile ones—and making the positive choice certainly has not marked human history so far. We will consider both sets of choices, but the alluring bright prospects, "green" on the dial, will be raised largely to show why we must continue discovery at this rapid pace. The thrust of discourse must concern the dangerous options, because that is where we must keep control, or win it, to ensure a civilized future.

The choices are not as simple as they seem. The modern drug-discovery process can be turned on its head to design "better" chemical and biological weapons. In terms of what that might lead to, no potential development inspires such universal fear and opposition as the hostile use of chemicals for mind control. But no other pharmacological product *sells* to such a legion of faithful users. Americans pop pills to brighten mood, dampen impulse, intensify senses. Yet just change the words "prescribing physician" to "deploying unit" to appreciate the sinister implications.

Right now, government-funded research to understand neurobiology and immunology, and pharmaceutical company development of new painkillers, anesthetics, antidepressants, sleep aids, allergy drugs, and antiinflammatories—all carried out for good reasons—pose the ultimate threat of hijacking our bodies and minds.

Imagine a world in which we can chemically induce every mental state at will and in whole populations—fear or fearlessness, paranoia or trust, false memory or forgetfulness. Medical journals affirm that we inch toward that day, their dazzling brain scans showing the compliance of each critical brain region to impelling drugs. Now listen to a Department of Defense advisory board: "Applications of biological, chemical, or electromagnetic radiation effects *on humans* should be pursued. R&D into sophisticated psychological operations designed to change the minds of individuals or the populace is needed."[1]

Needed? If our government needs to conduct research to chemically alter human mental states and emotions against a subject's will, we are nearing that worst outcome, and approaching it knowingly. The accelerating pace of the new biology may make it possible—all the more so if we do nothing to stop the military from pursuing this "need."[2]

Matthew Meselson,[3] an important catalyst in persuading President Nixon to move forcefully against biological weapons, argues that the hostile exploitation of these new discoveries "would distort the accelerating revolution in biotechnology in ways that would vitiate its vast potential for beneficial application and could have inimical consequences for the course of civilization." But doing just that is our history, he warns. "Every major technology—metallurgy, explosives, internal combustion, aviation, electronics, nuclear energy—has been intensively exploited, not only for peaceful purposes but also for hostile ones."[4] Meselson, a major science voice for more than forty years in efforts to quell biological and chemical weapons development, directs the Harvard Sussex Program on Chemical and Biological Weapons Limitation at Harvard's John F. Kennedy School of Government and the University of Sussex in the United Kingdom.

No one should doubt the horrific consequences such development might entail. Perhaps the highest profile group to raise the alarm is the International Committee for the Red Cross, which has issued an "Appeal on Biotechnology, Weapons and Humanity."[5] According to Robin Coupland, its medical advisor, the appeal "is not a trivial undertaking for the ICRC"—an understatement. He points out that nearly a century has passed since the committee made its last public appeal in the field in 1918. That appeal "eventually helped to bring about the 1925 Geneva Protocol," prohibiting use of chemical and biological weapons in warfare.[6]

Horns of the Dilemma

Pharmaceutical industry fermenters stretching several stories skyward breed enormous quantities of *Streptomyces* bacteria, a major source of lifesaving antibiotics. Identical fermenters could as easily grow huge batches of anthrax or another microbe that might be a candidate for a bioweapon. Fermenters are just one item of equipment termed "dual use," not because they are now used for both benign and destructive purposes but because they could be, in many cases without modification. And there are plenty of others, such as milling machines and aerosol-generating equipment, all invented and used for beneficial ends, all capable of being diverted toward bioweaponry. A hostile nation that had turned its drug fermenters into anthrax breeders might mill the spores for delivery as clouds of fine powder that could be aerosolized.

Perhaps it is obvious that equipment can be put to so many unrelated purposes, but what about microbes themselves? There is still no commercial use for anthrax or plague bacteria or their toxins and other components. But there is for *Clostridium bo-*

tulinum, which has the ominous distinction of producing the deadliest toxin known, so lethal that a raindrop's worth has the power to kill thousands if it could be delivered perfectly. Tens of thousands of aging Americans know the cosmetic benefits of tiny amounts of botulin as the active ingredient in Botox, magical remover of frown lines and wrinkles. However, that does not tap the beneficial potential of the toxin in medicine, and later we'll look at the sound reasons for research to make botulin even more potent than it is in nature.

Waiting in the Wings

Some strategies of basic research and drug discovery and development could easily be turned toward bioweaponry with the technology already available. For example, one critical shield against bacterial weapons would be powerful antibiotics that could be administered either as a preventative or to eliminate existing infections. But as we know too well, bacteria have developed resistance to many of our best antibiotics, and some of them have developed multiple-drug resistance, or MDR. Anyone who has been infected by an MDR bacterium knows that getting rid of its potentially life-threatening infection can be both difficult and expensive—and it might not work at all. Before antibiotics were discovered, over 50 percent of victims of all staph infections died,[7] and mortality rose to 82 percent of those with staph blood infections.[8] There is fear that we may again experience these pre-antibiotic mortality rates, this time from MDR varieties.

One bioweaponeering strategy for improving the killing power of an agent would be to isolate the source of MDR in a resistant strain of a fairly benign bacterium such as *Escherichia coli,* a ubiquitous resident everywhere from the human digestive system to lakes and ponds, or in a resistant strain related to anthrax, and then engineer it into their own bacterium. This is easier than it sounds and has been achieved by at least one offensive bioweapons program—the massive one developed by the Soviets—for a few bacterial agents.[9] The genetically conferred resistance usually resides in small, circular pieces of DNA called plasmids that are separate from the bacterium's genome, its main DNA complement. Plasmids have been basic genetic engineering tools for decades, and they are easy to isolate. With a few routine manipulations by a skilled molecular biologist in a well-equipped laboratory, a plasmid with resistance genes to several antibiotics can be inserted into the bioweapon agent and replicated, leaving the lethal agent *itself* defended against a range of antibiotics and leaving victims with fewer antibiotic options.

Decades of development of genetic engineering tools have enabled sophisticated molecular biology laboratories to unlock the DNA sequence of the human genome and those of a host of other organisms, including many pathogens and bioweapons agents. Anyone can download the DNA sequence of smallpox from the Internet.[10] Right now, that is like having complete pictures of a weapon and its parts—but not understanding the specialized fasteners or having the intricate assembly instructions. However, with the increasing ability to synthesize long pieces of DNA, scientists should soon be able to build an entire smallpox genome. Someone may even be up to it right now. Luckily, there is still a distance from there to a potential bioweapon: it would be difficult and require a sophisticated molecular virology lab to enable such a "handmade" smallpox genome to generate a live, infective virus. Yet even that feat has already been accomplished with the much smaller poliovirus,[11] so the gap likely will be closed.[12]

On, now, to botulin and the host of protein toxins that are crafted by bacteria and other microbes for their own purposes but cause humans and other animals sickness and sometimes death. That's so even in their natural state, but all proteins have their blueprints and assembly instructions embedded in their genes. That means by tinkering with those genes, genetic engineers might create a panoply of "new and improved" toxins.

Such toxins are already familiar in different contexts. The venoms of snakes, insects, and spiders; plant toxins, such as the ricin in castor beans; and bacterial toxins produced by cholera, diphtheria, and tetanus are all proteins, and they are all able, as biosecurity expert Jonathan Tucker notes, "to incapacitate or kill a human at remarkably low doses."[13]

The toxin ricin is already categorized as a potential biochemical weapons agent, and it is one that is fairly easy for a technician skilled in protein isolation to acquire. Clearly, ricin genetically engineered to be more potent would be a larger threat, but this is not a route that terrorists lacking a sophisticated molecular biology lab would find promising.

Delivering toxins on a mass scale would present an enormous problem, even for a warring nation, but brewing them in huge quantities would not be difficult. The DNA sequences of many such genes are known. Using standard techniques of molecular biology and biochemical engineering and a traditional biology laboratory workhorse such as *E. coli,* many toxins could be produced in huge quantities in industrial fermenters. Other techniques would enable enemy scientists to prepare and test large numbers of mutant protein toxins to search for one that is deadlier or resistant to an existing treatment, or even one that might escape standard detection tests.[14]

All of these, of course, represent the "evil twin" of dual use. However, on the bright side, natural villains can be turned to a whole range of beneficial uses, and removing the wrinkles of aging is the least of them. The protein toxins zero in with great power and specificity on their molecular targets, so they are of great use in basic research.[15] Chemistry professor Kim Janda and his team at Scripps Research Institute in La Jolla, California, have been working with that deadliest botulin to boost its killing power even more, so far multiplying it fourteenfold. Their goal: to improve what is already a useful tool for treating multiple sclerosis, stroke, cerebral palsy, migraine, and even backache. They have found small-molecule activators of the toxin that permit using it in lower doses that therefore should dampen patients' harmful immune response to botulin.[16]

The work to enhance the toxicity of botulin perfectly illustrates why decisions to carry out dangerous experiments and to publish results once they are known must be made case by case. Botulin already is more than deadly enough; the problem for an enemy would lie in weaponization and delivery. As a result, these experiments and their publication did not increase our risk—unlike the resurrection of the 1918 pandemic flu virus and its subsequent study in so many labs, which we view as the most dangerous current activity—and one that should not be continued, as we will discuss later.

All these strategies to weaponize the products of the new biology could be exploited by any nation with routine molecular biology and genetic engineering capabilities, though some work would require more sophistication. They could be used by terrorists only if they are able to buy or steal or are given pathogens and other required material from a state-run program. The bottom line is that we must bend our efforts to prevent the hostile exploitation of biology by nations, and that requires actions that begin on our doorsteps and extend everywhere.

World Enough and Time

In one modern arena all the tortuous pathways from laboratory discovery to usable molecule come together, a playing field where every day the struggle to turn incontrovertible test-tube results into runaway success plays out with scoreboard results as plain as black and white—or rather, black and red. In the pharmaceutical industry, betting on winners and ferrying them through rough waters to market means live or die.

Drug discovery and development tells much about potential problems to be expected in the pursuit of advanced biological weapons. How new drugs come to market, how most fail, and the years of painstaking endeavor required for a win *or* a loss offer a remarkable window into the vast gap between idea and execution. Pharmaceutical companies are the biggest spenders on the knowledge pouring out of the new biology, because they are its biggest users. That's one way they hedge their bets.

Discovery and development represent two different phases in the process. The first phase: find a molecular target that is a key link in a disease process, then identify a leading drug candidate that will in some way interfere with its action. In the development phase, that lead candidate is taken first into preclinical trials in animals for safety and efficacy testing. Then come the years-long, arduous, three-phase human clinical trials.

If drug discovery and development is becoming efficient and routine, that might forecast a boom in advanced bioweapons. Here are the standings in the gigantic global industry known as "pharma": thirty years ago, before most of the advances in the new biology, only 20 percent of drugs entering clinical trials made it to market, having traversed the complex and very costly path through clinical trials to ascertain their safety and effectiveness. Today, no change: only 20 percent make it to market.

Meanwhile, bringing that new drug through discovery and development and to the pharmacy and physician's office cost an average $802 million per successful drug in 2003 when the last formal study was completed, compared to a few hundred million dollars just ten years earlier.[17] About a quarter of that cost is for discovery, and total cost increases substantially every year, so it is now likely more than $1 billion.[18] That kind of price tag for a radically new bioweapon discovery would be beyond the range of all but developed nations with at least the financial wherewithal of large drug companies, many of which are as wealthy as small countries.

Why so little change in an era of such progress? Here's an educated guess. Nature has myriad complexities that we are just beginning to realize. What you see in the lab is not what you get into the pharmacy—or the armory. For a glimpse of these complexities, we need only look at the Human Genome Project.

Changes in drug development time and cost so far have not come with biological advance, but change clearly has come to the armamentarium—the medical term encompassing all the equipment, pharmaceuticals, and methods used in medicine.

Stories of Self

The Human Genome Project set one of the most ambitious goals in the history of science: to understand the "book" of instructions that makes us human, as well as the books for other living things ranging from the ubiquitous bacterium *E. coli* and brewers yeast, *Saccharomyces cerevisiae,* to the deadly smallpox virus, *Variola major.* As even high school biology students now learn, the instructions for humans to reproduce, grow, and survive in the world are three billion "letters," assembled in combinations of the four-letter alphabet of DNA, written as A, G, T, and C. The letters are short for adenine, guanine, thymine, and cytosine, the DNA chemical building blocks.

Project scientists now have written down the entire sequence, enough to fill a six thousand–book library. A piece of one gene might read AGGTCTAGGC, and the full gene might specify part of a structural protein of a cell, or of an enzyme to facilitate and empower actions in the cell or the entire body, or it might specify part of an instruction for the genes themselves. It is a wondrously simple scheme that was incredibly difficult to decipher for a genome as huge as ours, an epic spanning nearly fifteen years and the efforts of thousands of scientists, a few of whom garnered Nobel prizes along the way.

But as a measure of the accelerating progress in biology, during those fifteen years the technology to do the job improved so greatly that if it had been available from the start, the project would have ended almost as it began. In a showboating 2008 demonstration, the entire genome of James D. Watson [19] was sequenced—not in fifteen years but in two months.[20]

That, in turn, led to another leap forward, and the long-awaited $1,000 genome may be at hand. Soon—if not by the time you read this—if you submit your DNA for analysis, your genome will not have to be deciphered from scratch, as was Watson's. Ultra-fast "re-sequencing" methods can simply isolate differences between your genes and those already compiled into an "average" genome. Next: wearable microchips, unique to each of our six thousand volumes of instructions, including those for twenty-five thousand individually tailored genes.

Joining Forces

The Human Genome Project has both contributed to the emergence of the new biology and been aided by it. The new biology itself represents not the consequence of a singular trailblazing discovery—a theory of evolution or relativity, or finding the structure of DNA. It emerges from the resonance of many rapidly emerging disciplines. Those wearable microchips should be a landmark in the quest for individualized medicine—drugs and therapies tailored to each of us instead of one-size-fits-all with the concomitant side effects.

Today many of the methods of modern drug discovery feature massively parallel experiments, with thousands carried out at once instead of serially. Researchers can screen thousands of proteins as potential drug targets and a like number of potential drugs at the same time. The concept is illustrated in figure 2.1, where the significant shortening of experimental time becomes obvious.

The most striking example of massively parallel bioprocessing is the DNA microarray. Consider that this device can analyze all the genes in a human cell in one experiment lasting weeks. By contrast it might take fifty *years* to examine the twenty-five thousand human genes a few at a time. You might imagine this powerful tool as an array of heavy-industry proportions, but it is the size of a postage stamp, and the whole device into which it fits is smaller than a video iPod (see figure 2.2).

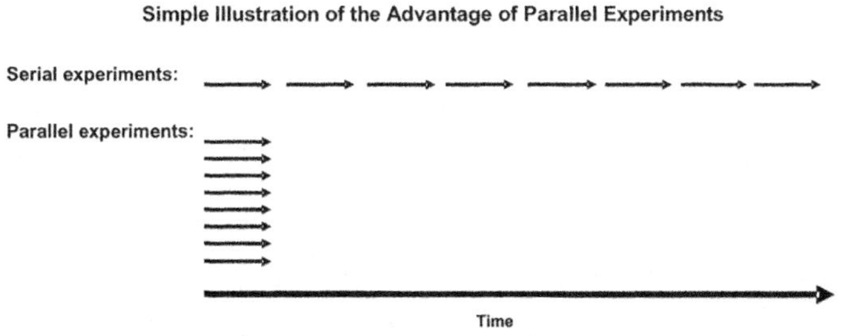

FIGURE 2.1. Illustration of the time difference to complete serial and parallel experiments

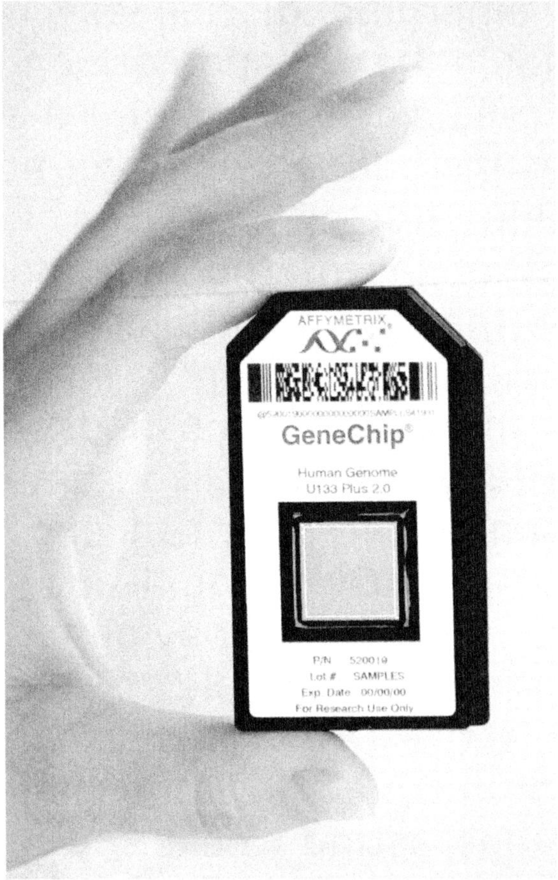

FIGURE 2.2. Microarrays, such as Affymetrix's Genechip (above), now include all known human genes. (Courtesy of Affymetrix.)

The discovery process is identical whether you are hunting for a miracle drug or a radically new chemical weapon. In both cases you are trying to identify small molecules—on the beneficial side, those that fight disease, on the malevolent side, those that are in some way highly lethal. Your search is among the twenty-five thousand human genes. In one case you look for a gene whose protein product is a disease target, in the other for a gene whose product is critical to life itself, so disabling it would kill us. Then you generate thousands of small-molecule candidates that bind to the respective target proteins to shut them down. Great success means finding just the right match. Green: wonder drug. Red: potent new chemical weapon. Clearly, the rapid speedup of the discovery process offers a potential boon to bioweapons developers—especially if they were trained among the new breed of engineers.

Armamentarium to Armory: Military Exploitation

When we learn the reason nearly everyone's potential wonder drug devastates a minority of patients, we may also learn a terrible secret: that that minority of patients is within a specific ethnic group. As readily as we might cure members of that group with specially targeted drugs, we might attack them with specially targeted weapons. There is no "lady or tiger" here; opening one door using such new techniques admits both.

Fortunately, sharp boundaries for ethnic sensitivities to drugs have not been observed, and there is considerable genetic variation within ethnic groups, so ethnic weapons remain only a theoretical possibility—for the time being. The moral implications of the development and use of such weapons are obvious, but what do you, the discoverer, do if the drug has lifesaving value for most ethnic groups but is harmful in one? Do you develop and market the drug with that warning, or do you quietly bury it?

Recently the Russian newspaper *Kommersant* reported that President Vladimir Putin banned export of human biological specimens from the country.[21] According to one unnamed government official quoted in the story, "Several large Western medical centers that receive shipments of biological materials from Russia are said to be involved in the development of 'genetically engineered biological weapons' for use against the Russian population." While an ethnic weapon that targets Russians is far-fetched given the ethnic diversity within Russia's population, such an extreme measure shows the level of government concern over the possibility.[22] But biological specimens play a key role in understanding genetic and other diseases. Banning their export from Russia could slow important medical research that would benefit Russians as well as the rest of the world.[23]

There is another military side to the very different susceptibilities individuals have to drugs and other chemicals, and that relates to incapacitating agents. Age, health, and a host of other factors come into play when the military seeks to use drugs as nonlethal weapons, and those factors are far more random and hard to control.

Controls on chemical and biological agents have evolved over nearly a century of complex international agreements. The Geneva Protocol of 1925 banned using lethal chemical and biological agents *in warfare.*

Later treaties separated rules governing biological and chemical agents, generally increasing restrictions, but here it is worth looking at how tortuous the rules for chemical agents have become. The Chemical Weapons Convention added a ban on development, production, and stockpiling—not merely the use in warfare—of lethal chemical agents. But nonlethal agents, those designed to pacify or incapacitate rather than kill, were banned for use in warfare but *not* for domestic purposes.[24] The domestic exemption allows using nonlethal chemical agents for "law enforcement including domestic riot control purposes." By extension, this could legalize the use of nonlethal agents in military-occupied territories, where occupying forces are required to enforce the law. That is the resolution the U.S. military is driving for.

The BWC makes no such law enforcement distinctions, so biological incapacitating agents would be illegal under any circumstances.

As one general put it succinctly, the military needs something between "a bullet and a bullhorn." Such incapacitating agents would save lives while achieving military goals, or so the argument goes. Unlike tear gas or pepper spray, the military seeks something that truly incapacitates so the target cannot flee, fire a gun, or put up any other kind of resistance. Would it not be more humane to flush out enemy combatants in order to capture rather than kill them? *New York Times* reporter William Broad expressed the humane-agent sentiment in an article whose headline captures the message: "Oh, What a Lovely War, If No One Dies."[25]

The framers of the CWC already had the answer to that argument, taken from recent American history. During the Vietnam War, tear gas was used to flush out enemy fighters *in order* to kill them, so the convention included a ban on tear gas and other such agents in warfare.

But let's focus on the use of "nonlethal" agents for a determinedly nonlethal purpose: complete incapacitation. That was the goal of Russian authorities in ending the Moscow theater siege by pumping in a mist purportedly containing fentanyl, a very commonly used hospital anesthetic and chronic-pain reliever. Fentanyl also has the distinction of being a major drug of abuse, nicknamed "gentleman's heroin" among addicts. The dousing killed all 39 Chechen insurgents, but it also killed—immediately or soon after—as many as 129 out of 800 hostages.[26] The percentage killed was higher than that of the mustard gas remembered as such an awful weapon of death from World War I—although gas's reputation was as much due to the agonizing death it brought on as its kill rate. Nevertheless, the Moscow victims did not all die painlessly in the siege. Some died long afterward, many others remain maimed for life, and many suffer from illnesses directly related to their exposure.[27]

The lessons? As in so many other muddied modern conflicts, opponents and proponents each see their side supported. In an interview with Wired.com , retired Green Beret Colonel John Alexander, the most vocal proponent of nonlethal weapons, cited their use in the siege as "a good example" of their proper role.[28] When the interviewer noted the high death toll, Alexander pointed out that six hundred were saved from attackers who had already shot several hostages, and he expressed certainty that the rebels were prepared for everyone to die.[29]

Alexander could be right in this case, but it is probably impossible to design any incapacitating agents that are nonlethal. To legalize their use might be tantamount to licensing killing, and continuing their use could lead to death tolls far exceeding that in Moscow. The reasons owe to the demands on commanders in deadly confrontations and to other circumstances beyond anyone's control.

In their article "Beware the Siren's Song," scientists from the Center for Arms Control and Non-Proliferation cite a variety of reasons that a high kill-rate must be expected from such agents.[30] Several of the more critical reasons follow:

- Given the requirement that the weapons must incapacitate about 99 percent of combatants to achieve tactical goals, high concentrations are a must.
- The agents must act not only surely, but quickly—in less than a minute—before targets can react. That demands still higher doses.
- Aerosols pumped into the enclosed but public spaces typical in military-insurgent encounters

continue to be inhaled by victims until they are evacuated, often long after they are unconscious. That can result in doses many times those planned, and with fentanyl, death from cessation of breathing.

- For the military commander, the potential costs of using too much agent and increasing mortality are far lower than of using too little and risking losing his own forces or the battle itself.
- Even more seriously, the very young, old, or sickly—almost certain to be represented among hostages and innocent bystanders—are far more likely to die from any given dose than the young, healthy combatants who are the targets.

In recent years, even proponents have taken to calling these "less-thanlethal agents" or "calmatives" in hopes of damping public opposition to them, but our objections are firmly grounded in both pharmacological and theoretical arguments.

Can fentanyl be made safer or even truly nonlethal? Breathing cessation is a fatal side effect of the drug and the likely cause of many if not most of the immediate deaths in Moscow. Given the value of fentanyl as an anesthetic, there might be a clinical benefit to making it safer, although during surgery breathing can be maintained on a respirator.

In a 2001 study, German researchers showed that in animals under fentanyl anesthesia, breathing could be maintained by using a chemical called BIMU8 that stimulates the breathing reflex.[31] BIMU8 is a highly complex chemical, as its unwieldy chemical name implies: endo-N-8-methyl-8azabicyclo[3.2.1]oct-3-yl)-2,3-dehydro-2-oxo-3-(prop-2-yl)-1H-benzimidazole-1-carboxamide.[32]

Could this complex chemical be included in a fentanyl mist? Maybe. Would BIMU8 reach its target to stimulate breathing? Maybe. But maybe is not likely. It is unlikely that something off the shelf will work in the field. It is unlikely that using fentanyl outside the tight controls of the operating room will prove safe. And even if fentanyl victims keep breathing, other fatal effects might well appear, effects that were masked in Moscow by rapid suffocation. That seems likely because of the incredibly high concentrations of fentanyl needed for rapid, complete incapacitation. However, the German paper probably holds enough promise to spur on the military.

An excellent review of the history, present status, and technical and strategic aspects of military development of incapacitating agents may be found in a recent report by the Bradford Disarmament Research Centre, curiously titled "'Off the Rocker' and 'On the Floor': The Continued Development of Biochemical Incapacitating Weapons."[33] For many years, the Bradford group has led the opposition to development and use of incapacitating agents. The report points out that the sites of action of many potential incapacitating agents are known,[34] and coupled with advances in genomics and neurobiology, the time may be right for the development of better incapacitating agents, although safety and delivery still are serious issues.[35]

The Bradford group is not alone in its concerns. Julian Robinson and Matthew Meselson, two longtime opponents of military use of nonlethal chemical weapons, write, "It is hard to think of any issue having as much potential for jeopardizing the long term future of the Chemical and Biological Weapons conventions as does the interest in creating special exemptions for so-called non-lethal chemical weapons." The reason is basic: "Exemption blurs the simple line, *no poisons in war.*"[36] At the urging of Meselson, the nonpartisan Council on Foreign Relations came out against non-lethal chemical weapons in warfare because their use would erode the both the Chemical and Biological Weapons conventions.[37]

Well into the writing of this book, the prestigious British Medical Association came out forcefully against any use of nonlethal chemical agents as weapons.[38] In its report, entitled "The Use of Drugs as Weapons," the association takes positions exactly in line with ours on the true lethality of nonlethal chemical agents, relying frequently on arguments from "Beware the Siren's Song." It also takes positions in line with ours on subjects ranging from concerns over the pace of modern biotechnology to arms control. The report also reminds physicians of their ethical responsibilities.

Improving the safety of existing incapacitating agents remains a possibility for the near future, but designing new mind-control or incapacitating agents is, in our opinion, beyond the reach of present-day biology, as we will soon discuss.

Agents of the Cold War

The story of the government's search for ways to influence thinking or to incapacitate began in agricultural fields in the 1950s. For many cold war years later, the CIA conducted experiments with the derivative of an alkaloid found in wheat ergot called LSD, expressly for its possible uses in interrogation and mind control.[39]

In 1977, Senator Ted Kennedy reported to the Senate, "The Deputy Director of the CIA revealed that over thirty universities and institutions were involved in an 'extensive testing and experimentation' program which included covert drug tests on unwitting citizens 'at all social levels, high and low, native Americans and foreign.' Several of these tests involved the administration of LSD to 'unwitting subjects in social situations.' At least one death ... resulted from these activities. The Agency itself acknowledged that these tests made little scientific sense."[40]

At the same time, another U.S. program not related to the CIA's, this one at the Edgewood Arsenal in Maryland, was conducting experiments on agents including BZ, one of the tranquilizing benzodiazepines;[41] LSD; and synthetic marijuana to disable enemy troops. James Ketchum, a colonel and key scientist, has written an entertaining book entitled *Chemical Warfare: Secrets Almost Forgotten—A Personal Story of Medical Testing of Army Volunteers with Incapacitating Chemical Agents during the Cold War (1955–1975).*[42] Unlike the CIA program, research subjects all signed informedconsent forms, both a general one and another related to any experiment they were to participate in.[43] Experiments were carried out with safety of subjects a principal focus, Ketchum said. He wrote:

> Over a period of 20 years, more than 7,000 volunteers spent an estimated total of 14,000 months at Edgewood Arsenal. To my knowledge, not one of them died or suffered a serious illness or permanent injury. That adds up to 1,167 man-years of survival. Statistically, at least one out of a thousand young soldiers chosen at random might be expected to expire during any one-year period. By this logic, Edgewood was possibly the safest military place in the world to spend two months![44]

This would seem to be a powerful argument for the safety of incapacitating agents. But it is precisely because these experiments were conducted with safety of the subjects as a priority that they offer little evidence of how safe incapacitating agents would be under conditions like those during the Moscow theater siege. At Edgewood, even at the highest doses it often took an hour or more for incapacitating effects to show, and the end-effects usually did not include full incapacitation, let alone unconsciousness. After all, the Edgewood experimenters were focused on disabling soldiers in combat, where there would be tactical value simply in disabling the enemy.

For this and other reasons, we and Ketchum remain far apart in judging the safety of using the Edgewood or any purportedly nonlethal agents. The only way to resolve our differences would be to experiment under Moscow-like conditions using BZ, LSD, or some other incapacitating agent at concentrations that would meet a riot-control commander's requirements, that is, fully incapacitating attackers—and anyone in their midst—within a minute or two. And we would never sign on to that experiment, for the obvious reason that it would likely kill at least some of the experimental subjects. Even the military would not sign on.

Nonlethal agents can be most lethal. That was the lesson from Moscow.

Hijacking Our Minds and Bodies

There is, of course, a whole range of potential agents aimed at guaranteeing nonlethal pacification. "R&D into sophisticated psychological operations designed to change the minds of individuals or the populace is needed," as the Defense Department advisory board put it. What once were faltering efforts to bend and persuade through subliminal messages and disinformation might be transformed by chemicals derived from the explosive advances in neuroscience.

Spraying a captured city with a drug that would make its residents trust their new masters or make its insurgents throw roses at them may seem outlandish, and extremes in behavioral control are probably years off. But in 2005, Swiss researchers reported in the journal *Nature* that test subjects given a nasal spray of the natural brain hormone oxytocin showed markedly increased trust in those with whom they were dealing in social situations, compared with those given placebos.[45] The drug lowered their defensive posture toward those offering them financial deals in the laboratory experiment. That and other research offers strong evidence that in fact this hormone, known to be a player in maternal bonding at childbirth, enhances human social bonding in general, even when we might reasonably remain skeptical or aloof.

The U.S. company Vero Labs already is marketing oxytocin as a drug to render its wearer trusted. In fact, Liquid Trust is the name, and Vero claims it is "the world's first and only product to attract women by getting them to trust you." Says the manufacturer on its Web site: "Made with pure Oxytocin—A hormone that is scientifically proven to make women trust you more." That's just the pitch aimed at singles, replete with images of couples kissing on the beach, and of a man having his tie removed next to a bed by a woman clad in a teddy. Moving on to the executive suite and sales meeting, there are business suit–clad men clasping hands, people in convivial gatherings—presumably only one of them wearing Liquid Trust.[46]

These claims may be wildly exaggerated, but it is worth considering that government agencies for decades have been quietly at work on the "problem" of forcing trust under questioning and creating social bonding that runs counter to individual preferences. It would be naive to believe governments are not investigating every discovery concerning these issues for a potential application.

The noted neuroscience researcher Antonio Damasio [47] observed in an article accompanying the *Nature* research that sophisticated marketing and persuasive techniques already long used in sales and politics may actually work by stimulating the natural release of oxytocin in shoppers and voters. "Some may worry about the prospect that political operators will generously spray the crowd with oxytocin at rallies of their candidates," he wrote. However, he concluded, "Civic alarm over such abuses should have started long before this study."[48]

Precisely. Indeed, warning signs have been present as the years have passed since the *Nature* publication, but they fall on deaf ears. Bioethicist Jonathan Moreno sounded the alarm this way: "Neuroscience, neuropharmacology, and various novel devices are being touted as presenting significant advances in some areas of interest to national security agencies ...[T]he very process of exploring them raises fascinating, difficult and sometimes disturbing questions for social ethics and public policy."[49] And Vero Labs has marketed its Liquid Trust without a protest.

Hijacking our bodies: as we know from the Australian mousepox experiments referred to in the opening chapter, some proteins delivered by a virus can suppress the immune response, preventing a reaction to the virus's own alien presence. In the Australian experiments, a gene that was delivered via a mild mousepox virus accidentally suppressed the mouse's immune response. That turned the "mousy" virus into a powerful mouse killer. Immune-response suppression might be engineered into either viral or bacterial weapons agents, damping or destroying our immune systems' ability to fight back and leaving us defenseless against even ordinary pathogens. Molecular biologists still don't have a complete understanding of the intricacies of controlling the immune system, but, just as the mousepox experimenters accidentally discovered a lethal immune disabler, there is always a danger a researcher may set out using a hit or miss strategy to deliberately discover such a weapon.

Some scientists, however, are skeptical that we can create bioweapons more efficient than natural disease. David Baltimore, who won a Nobel Prize in science in 1975, notes, "The real danger today is from organisms that already exist. The idea of synthesizing something worse than that, of taking bits of Ebola and other viruses to create something more deadly, underestimates how hard it is to survive in the natural world. Adapting to the human lifestyle is very complicated, so I would guess that we would fail if we tried to engineer a dangerous organism."[50]

However, Moreno cautions us that the Defense Advanced Research Projects Agency (DARPA), which is now heavily into neuroscience research, has an impressive history of success in cutting-edge science: "The secret of DARPA's success is not its funding ... but its brilliant use of intellectual capital." To make his point he says, "DARPA designed the computer mouse and, to give the mouse something to click for, ... the internet, first called the DARPAnet *[sic]*."[51]

Are We There Yet?

What stands between us and mental subjugation? There is that enormous gap between the idea and its execution. Consider the following as one metaphor for the complexity of the human mind. The self, to a scientist, emerges from one human brain with its hundred billion neurons, or "thinking cells," each with so many thousand connections to others that numbers in neuroscience begin to approach those of stars in the universe, and there are no down-to-earth comparisons. At least that was the "classical" picture before things got complicated. Now scientists believe that the brain cells thought only to support the neurons, which are an order of magnitude more numerous, actually are key players in making the brain work. Controlling all that "wet-ware" in people is not a near-term fear.

These observations on brain complexity pertain to designer drugs constructed from the knowledge and methods of the new biology. But consider LSD, date-rape drugs, and opiates—and their ability to control people—all of which were discovered without understanding the complexity of the brain. Not quite mental subjugation, but nonetheless often quite effective control agents.

Then there is the nature of science itself, which always seeks the simplest possible explanation for all the critical evidence found in unraveling any conundrum, but which has a way of finding ever new mind-numbing complications. So it is with the marvelous ways that the genes encoded in every cell's DNA drive the manufacture of all the proteins that make an organism, from structure to function. The half-century-old understanding of how this proceeds step by step was based on what is called "the central dogma of molecular biology," a wonder of simplicity. A wonder undone by still greater mysteries.

Breakdown

There is a much larger scientific obstacle to surmount than mountains of data before we understand human health and disease at the molecular level, and that is the loss of the paradigm that gave scientists an anchor.

According to the central dogma, one gene encodes one messenger RNA, which then carries the instructions for making one protein out of the nucleus and onto the cell's "workbench." That would mean twenty-five thousand proteins for humans—a large number but manageable. Now scientists have found that each gene might give rise to several different messenger RNAs, each of which might yield different proteins. It is now estimated that humans actually make a million different proteins with that suddenly small number of genes. That was complicating enough, but then they found that even a single protein operates in consort with other proteins and with DNA itself—that is, a protein might function quite differently, depending on the company it keeps. If that sounds as complex as social behavior, it just might be.

Now we have moved from the fundamental and linear ABCs that can be encompassed in a kindergarten song directly into great literature, with no primer in between, where context and word order, emphasis and nuance are everything. Back to the very real fears for a future that would be both out of control and over-controlled. Are we there yet? No, fortunately. The enormous complexity of the living world that hinders our advance in medicine also hinders the fearful aspects of biological warfare. Now is the time to develop biology for our benefit and not our destruction. The enormous difficulty in achieving both productive and malign goals costs us progress, to be sure—and buys us time.

CHAPTER THREE

A Hawk Turns to Peace, Doves Go to War

> The introduction of radically cheap weapons of mass destruction into the arsenals of the world would ... endow dozens of relatively weak countries with great destructive capability. Such weapons could even come within the reach of dissident private groups and individuals. It is obviously to the advantage of great powers to keep war very expensive.
>
> MATTHEW MESELSON

With the 1925 Geneva Protocol, the use of biological and chemical weapons in warfare theoretically ended—but only for those who had ratified it. Japan had not. Warring against China in World War II, a Japanese military unit led by a physician-soldier secretly violated virtually every prohibition, amply demonstrating why a global agreement against these awful weapons was desperately needed to avert a drift to disaster.

But it was only decades later that a determined hawk took the American presidency and abruptly reversed the world's course, at least for a time.

It was 1969, a year of social turmoil at home and a seemingly endless war in Vietnam that was spreading into neighboring countries just as hawks and doves had predicted, though for different reasons. In major American cities, streets filled with protesters or rioters and the smell of CS tear gas often choked the air. The far left called for national revolution, and regardless of how unlikely such a massive upheaval might seem in hindsight, the notion then did not appear far-fetched.

The Russians were sparking revolutions around the globe, fueling them with the latest weapons from their arsenal and sending advisors to turbulent countries from South America to Asia, from the Middle East to Africa, just as the United States armed counterinsurgents and allies. Who knew what was in those Russian arsenals? Fears of nuclear holocaust were based on certain knowledge, but what about biological and chemical weapons? That was a black box, not only to the public but, as it turned out, to the CIA.

In the face of such dangers, President Richard Nixon made a decision that must have startled political friends and foes alike. He radically changed his mind about continuing to develop, produce, and stockpile biological weapons, ending an extensive decades-long program.[1] Bioweapons had been in development for the entire twentieth century and had been used at least once, by the Japanese in World War II, as we will spell out in detail. After less than a year in office, the famously resolute Nixon declared that mankind "already carries in its hands too many seeds of its own destruction." He said biological warfare had "massive, unpredictable, and potentially uncontrollable consequences" and was "repugnant to the conscience of mankind."[2]

Nixon could not have better summarized the horrific history of biological and chemical warfare and the research that enabled it in the twentieth century, from the gas-choked trenches of France to the all but forgotten Japanese killing fields in China.

In a stroke, he shut down the U.S. offensive biological weapons program and set the world on the path that would end in a global treaty just six years later, as the Biological Weapons Convention became international law.

Privately, Nixon was frustrated with the limitations of arms control agreements. In just-released government documents from 1969 to 1976, in his often blunt and profane manner, Nixon refers to his signing the BWC: "I ... went over to sign that jackass treaty on biological warfare." Earlier he had explained his cynicism to Kissinger: "As far as these agreements are concerned, they are basically not an end in themselves.... They limit arms but they do not mean the end of war. They are means to an end and that end is peace."[3] On the other hand, his reference to "that jackass treaty" may simply be a reflection of this frustrations dealing with the obstinate Russians.

What led Nixon to reverse a course followed by a quarter century of his predecessors? The story that reached its powerful climax in 1969 began a few years earlier, when the person who would become a most critical link in the president's change of mind was a freshly minted assistant professor of biology at Harvard.

Matthew Meselson was about to play a pivotal role in U.S. bioweapons policy, in the process carving out a career path he'd never imagined. Today, as he greets a former colleague in his Harvard office opposite the library where Henry Kissinger sometimes held court, he is among an elite group of scientists occasionally advising a succession of administrations on national science policy.[4] He has a thoughtful and reserved demeanor, perhaps belying the relentlessness with which he has pursued an end to bioweapons through the administrations of eight presidents, through the banning of them by international accord, and through the present.

That course began in 1963, during the Kennedy years, when Paul Doty, longtime advisor to the government on nuclear weapons disarmament and a departmental colleague at Harvard, asked Meselson if he would like to spend the summer at the Arms Control and Disarmament Agency in Washington. He jumped at the opportunity, and why not? He thought "it would be a nice way to spend the summer" while advising the government in critical areas of his expertise.[5] But he got off to a rocky start.

His first assignment was to work on European nuclear affairs, a subject he knew little about, and he quickly went from uncomfortable to embarrassed as major figures trooped in to see him. Among them,

Paul Nitze, assistant secretary of defense for international security affairs: "All I remember is his fingers were bright yellow because he chain-smoked." Even Llewellyn Thompson, ambassador to the U.S.S.R. itself, "came to my little office." After a week, Meselson asked supervisor Franklin Long if he couldn't advise in areas closer to home for him: chemistry and biology.

Do anything you want, Long told him, adding, "We had a guy who did that. He got depressed and killed himself, so you can have his desk."[6]

Meselson said, "They had to send him to a special hospital, because he got depressed first and killed himself later. People who have high security clearance can't just go to any old place, so they go to a place where all the doctors and nurses have clearances."

On the bright side, the young scientist's officemate was the noted physicist Freeman Dyson, with whom he had studied quantum mechanics at Berkeley. Right off, Meselson decided to pay two important calls: on the government's bioweapons center at Fort Detrick, Maryland, to see what the United States was doing, and on the CIA to see what other countries were up to.

At the CIA, "it was clear we didn't know what the Soviets were doing. We knew they had some laboratories and we knew the location where there were military laboratories where they did biology, but we didn't know anything about weapons one way or the other."

On to Fort Detrick—and a flash of insight that would lead him all the way into the present. His guide was Leroy Fothergill, who had been there from the start. At the "big, creepy, tall building" that housed the giant, brewery-like fermenters used for anthrax production, Meselson asked, "Why do we do this?"

"It's a lot cheaper than nuclear weapons," Fothergill replied.

And the light went on. "Don't we want to make war so expensive no one can afford it but us?" he asked. Why would the United States want to pioneer a way to make war so cheap that virtually anyone could unleash a deadly attack? When he told Dyson his feelings, his former mentor said, "Sometimes our intuitions are wiser than we know." For Meselson, "My intuition was, this is a really dumb thing to do."

The first fruit of that insight was a long paper whose core was "this simple argument that we don't want to pioneer an ultra-cheap weapon of mass destruction that could give everyone else something that otherwise they would not have," especially when the United States already had megaton nuclear weapons.

Meselson was in the government for only those three short summer months, but he pursued the matter relentlessly through the succession of presidents. In a 1965 Pugwash conference paper he wrote, "Biological weapons possess a dangerous potential as a cheap means of inflicting massive civilian casualties. The potential of microorganisms and viruses as cheap casualty-producing agents arises from the extreme smallness of the amount which might suffice to infect the population of a large area."[7]

So far, Meselson had been careful not to widely publicize the idea that biological weapons were a cheap means of warfare. Two events led him to go public with his concerns. As the Vietnam War built up, Bertrand Russell accused America of practicing chemical warfare. Then he was asked to review a book by a former Fort Detrick director who argued "in this friendly, cheerful way" that biowarfare was very humane compared with ballistic weapons because, though deadly, it was less painful. But for Meselson, a critical difference was that, however painful they might be, bullets and shrapnel kill their targets. Bioweapons kill everyone.

In his 1964 review of *Tomorrow's Weapons, Chemical and Biological* by General Jacquard Rothschild for the *Bulletin of the Atomic Scientists,* Meselson wrote presciently: "The introduction of radically cheap weapons of mass destruction into the arsenals of the world would not act as much to strengthen the big powers as it would to endow dozens of relatively weak countries with great destructive capability. Such weapons could even come within the reach of dissident private groups and individuals. It is obviously to the advantage of great powers to keep war very expensive."[8]

The good news is that it is still very hard to build biological weapons and to deliver them successfully enough to cause mass death—for example, as a smokelike aerosol wafted over a city or a very fine powder mailed to hundreds of addresses. The bad news is that it would still be relatively inexpensive. Some argue that they may be within the reach of terrorists because of their low cost compared to nuclear weapons.

Meselson warned, "once this development is accomplished by any nation, the technology would almost certainly diffuse to others, in spite of security precautions."

Ironically, General Rothschild was an idealist who saw the final achievement of "tomorrow's weapons" as global peace under an umbrella world government that would enforce international disarmament and inspections. "Something of a surprise," Meselson says.[9]

The push for presidential action gained momentum. Milton Leitenberg, then a young Northeastern University professor, urged Meselson and the renowned Harvard biochemist John Edsall to get President Johnson's attention with a petition showing the solidarity of American scientists in ending chemical and biological weapons development.

Meselson holds up the massive, blue-bound volume with its single-page petition and page after page of signatures—five thousand in all—from Nobel laureates, heads of top academic departments, and leading scientists of all specialties. Noting the "enormous devastation and death" that would be inflicted on civilians, the petition reverberates with Meselson's warning that the weapons "could become far cheaper and easier to produce than nuclear weapons, thereby placing great mass destructive power within reach of nations not now possessing it."

But it went on to warn of seemingly innocuous developments that could lead to lethal outcomes. The reference was to the use of tear gas by U.S. forces in Vietnam, for riot control and to flush guerrillas from hiding, and of defoliants to destroy cover. It would be some thirty years before these two issues would be resolved, their use prohibited in warfare by the Chemical Weapons Convention in 1995. However, arguments supporting their use persist to this day, here and abroad. But at that point, nothing had been resolved, nor would it soon be. Indeed, the petition was reviewed by the president, but he in turn circulated it for review by affected agencies. The Department of Defense balked, and there the critical issue lay as administrations changed.

Then, as often happens, a chance encounter sparked a whole new chain of events. The scene was Boston's hectic Logan International Airport. Meselson almost collided with former colleague Henry Kissinger, one Washington-bound, the other headed for Cambridge. They embraced and Meselson congratulated Kissinger on his new job as national security advisor to Nixon. Kissinger responded, "What shall we do about *your* thing?"[10]

Kissinger knew well of Meselson's struggle against bioweapons development, and he was prepared to be the catalyst, bringing the scientist's reports to the new president's eyes. The papers Meselson wrote for Nixon did not force an opinion but set forth the options—the only way he thought a scientist ought to approach the chief executive. The decision would be Nixon's. But Kissinger was far more than a catalyst in the drive toward the policy change. Not one to let competing interests such as those in Defense sabotage a critical issue, this time when the documents went out to agency heads, Kissinger applied his considerable muscle until their reviews gave appropriate weight to the value of eliminating biological weapons.

"It was brilliant," Meselson says. "It reserved the option for the president, which is where it belongs."

Ambassador James Leonard, the U.S. negotiator of the Biological Weapons Convention, pointed out in an interview that an interesting confluence of events enabled Meselson to get leverage for the BWC. "The United States [was] under constant attack both internationally and at home here over the use of tear gas and herbicides in Vietnam," he recalled. "Nixon, Kissinger, and Melvin Laird, the defense secretary, all thought we ought to do something to get this monkey off our back—the whole problem of tear gas and herbicides."[11] The BWC emerged as that something.

Leonard says neither the Defense nor State departments would likely have pushed the convention forward on their own, "but when Meselson stimulated the machinery, then they swung into action and agreed." In the end, Leonard says, the State Department drafted the decision document.[12]

While it was the use of chemicals in Vietnam that provided the political impetus, ironically it was only biological weapons that Nixon renounced in his 1969 speech. However, there were political reasons for separating chemical from biological weapons. A year before Nixon's decision, Leonard says, the British urged:

> We ought to do a treaty about chemical and biological weapons. But that we should separate the two, chemical and biological, and do biological first, and put off chemical.... And our basic decision in November of'69 was that we would do BW without any verification.... When the experts in State and Defense and CIA were doing the studies in 1969, they simply reached the conclusion without any argument that a BW treaty would not be verifiable without extensive on site inspection. And we either had a choice of a treaty without any inspection, without any verification or no treaty. And they said lets get a treaty.[13]

America was in the middle of the cold war, so it was unlikely that the Soviet's would accept any kind of on site verification for chemical or biological weapons.

This separation was contrary to what most diplomats wanted, but "one by one they came over under our pressure and persuasion and reasoning."[14]

That much is history. What images lay behind Nixon's grim portrayal of biological warfare as fraught with "massive, unpredictable, and potentially uncontrollable consequences" and "repugnant to the conscience of mankind"? No one knows. But given his frequent invocation of history as prologue, it is possible he knew of the horrors visited on Chinese civilians and soldiers by Major Ishii Shiro, a physician as wedded to death as the notorious Josef Mengele.

"The dual thrill of discovery": Unit 731

Forty-six years before Nixon's declaration, the landmark Protocol for the Prohibition of the Use in War of Asphyxiating, Poisonous or Other Gases, and of Bacteriological Methods of Warfare was enacted. According to Ambassador Leonard, the document "was originally a protocol to the Versailles Treaty. And in 1919, many of the world's leading powers did gather in Paris or in Versailles to sign the Protocol but they couldn't quite do it." Finally, in 1925, the protocol became "a freestanding treaty."[15]

The Geneva Protocol of 1925 declares its prohibitions "binding alike the conscience and the practice of nations."[16] So why was further action needed half a century later? The devil hid in the details—development, producing, and stockpiling were *not* banned—only using the weapons in warfare. And many nations signed with the reservation that they would not be bound by the protocol if chemical or biological weapons were used against them.[17] "Those reservations effectively turned the Geneva Protocol into a no-first-use treaty," says Leonard.[18] And there were striking omissions from the list of countries bound by the protocol. Without ratifying it, countries were not "party to" the agreement, so it was not binding on them, and both the United States and Japan signed but did not ratify the protocol for decades.

The United States, fearing what other nations would do in secret, pressed forward to develop and stockpile the weapons. The irony was that so did other nations, including many that had ratified. However, although Allied nations and Germany and Italy developed and stockpiled chemical weapons in World War II, they did not use them, and the Geneva Protocol is credited for saving many lives.[19]

Japan was another story, and its central figure was Major Ishii Shiro. Unlike Mengele, who fled the Nuremberg war crimes trials and died in exile in hiding, when Ishii died of cancer in 1959, he was a wealthy man at home in Japan.[20] And unlike dozens of Nazi death merchants hanged or imprisoned, Ishii and his top officers were never indicted for war crimes, and investigation of them was suspended by the United States in return for their revealing the secrets of the human experimentation that the U.S. would not undertake as "repugnant to conscience." The three thousand men who worked for Ishii in his Unit 731 and the thousands more who worked in the many other "death factories" escaped as well.

Jeanne Guillemin, senior fellow in the Security Studies Program at Massachusetts Institute of Technology, cites security reasons for failure to hold the Japanese accountable.

> U.S. government agencies in charge of decisions were consistently motivated by national security interests, and in particular, the military option to pursue a biological weapons capability .... The U.S. Army's Chemical Corps, which housed the U.S. biological weapons program, had convinced the Joint Chiefs of Staff and other high-level Washington officials that information about the Japanese biological weapons program was vital to national security.[21]

And "The value to the U.S. of Japanese BW data is of such importance to national security as to far outweigh the value accruing from 'war crimes' prosecution."[22]

Martin Furmanski, who has extensively researched Japanese bioweapons development during the war, said, "In fact the Japanese data was worthless, but the U.S. BW program had an interest in preventing a Japanese BW war crimes trial." He speculated that those at the Fort Detrick headquarters of the U.S. bioweapons program "did not want the precedent of hanging a bunch of scientists for developing and using biological weapons!"[23]

During World War II, Japan stood alone as the only documented largescale user of biological weapons.[24] The Geneva Protocol convinced Ishii that disease must be a potent weapon or biological weapons would not have been banned. He lobbied vigorously and successfully for the Japanese to start an offensive bioweapons program.[25] Japanese proponents claimed that the Soviets were already engaged in extensive bioweapons research. And as with military strategists in other countries, the Japanese military believed in mutual deterrence. However, stories about Soviet bioweapons activities could have been disinformation by Ishii to convince the military and to justify an offensive program.

The largest of the dozen death factories and the base of operations for Ishii was a complex of 150 buildings where three thousand people worked on projects, including a pilot plant for making thirty-kilogram quantities of BW agents for use against Chinese civilians. Originally called the "Ishii Unit," it was renamed Unit 731, the name that lives on today.

Japan in effect turned occupied Manchuria into a biological and chemical warfare laboratory. Led by Ishii, soldiers and scientists experimented with and killed three thousand to ten thousand Chinese in their laboratories, where they were referred to as "logs"—to be experimented on and then studied via dissection, sometimes while still alive. In "field trials," ten times more Chinese may have been killed by plague, cholera, and other deadly bacteria spread in fields and drinking supplies.

"The plague/flea attacks against Chinese cities in 1940–1941 ... directly caused only a handful of cases," Furmanski said. "What these BW attacks did accomplish was the establishment of new endemic foci of plague, and in following years there were many hundreds and perhaps thousands of deaths. These larger outbreaks were more difficult to control, particularly due to the wartime conditions."[26]

Under Ishii's order, children were given chocolates laced with typhus or typhoid bacteria, and many died. The residents of Changchun, the capital of Manchukuo Province in Manchuria, were "inoculated" against cholera with a serum that actually contained one of those bacteria. An epidemic spread through the large city, giving Unit 731 scientists the opportunity to carry out epidemiological studies. Ishii experimented with every available means of delivering every variety of deadly organism in the most lethal ways.

On October 4, 1940 in the city of Quzhou, nine-year-old Qui Mingxuan watched as a single plane flew overhead dropping rags, soybeans, and wheat. The payload delivered bubonic plague germs and possibly cholera, typhoid, and anthrax bacteria. Soon thereafter, he recalled, "My relatives, friends, classmates have died."[27] Sixty years later, the bacteriologist remained especially bitter that "even families of the dead were not allowed to see their faces one last time at their funeral because they died at the isolation hospital."[28]

Qui charged that as many as fifty thousand people were known killed over a six-year period following that first outbreak of plague, which "had never existed in the town in all its history."[29]

American missionary Archie Crouch, "fresh from California" with his family in Ningbo, described one of Ishii's experiments in his diary. On October 27, 1940, instead of the usual bomber raid, a lone plane circled, trailing a plume of what appeared to be smoke. "I thought it must be on fire, but then the cloud dispersed downward quickly, like rain from a thunderhead on a summer day, and the plane flew away."[30]

A few days later, "the first bubonic plague symptoms appeared among people who lived in the center of the city." No one knows how many died in that ensuing epidemic.

But before anyone assumes this proved that plague could readily be weaponized today, Furmanski notes, "The war crime is clear, and the casualties were likely quite numerous, but one cannot draw the conclusion that because this weapon [plague] worked in 1940s wartime China that it would work in 21st century U.S.A."[31] For instance, antibiotics and public health measures can control plague, cholera, and other deadly bacterial diseases.

At least one major effort of Ishii's backfired. During a 1942 campaign, Unit 731 workers moved into an area vacated by their retreating army, dispersing a variety of biological agents to spread infection among the advancing Chinese. The attack boomeranged when Japanese troops inadvertently moved back in and suffered an estimated ten thousand casualties.[32] "Blowback" is the usual term for such disasters, originating with chemical weapons that blew back on the attackers. Warnings of blowback were used by the United States to convince other countries to become party to the Biological Weapons Convention.

Ambassador Leonard recalls, "We were arguing with great intensity and sincerity that biological weapons are just as much of a danger to the country that possesses them than they are to any adversary, the so-called blowback argument."[33]

For Ishii, the end of the war meant all was forgiven, or at least officially forgotten, as American victors realized they could have data available by no other means—and there was no way they would let the Russians have it.

Among the trove were thousands of tissue pathology slides that documented the extent of Japan's biological weapons research. But ultimately the findings proved of little value. The U.S. program focused mainly on delivering biological weapons as sophisticated aerosols in which the agent is stabilized for long-term storage. Other than simply spraying bacteria from an airplane, a primitive aerosol method, Ishii had used other means of delivery than aerosols, so his data were of relatively little use.

If there is one dark voice that should be heard by all scientists and administrators, by voters and activists contemplating how science can be coopted by politics, it is Ishii's. As Unit 731 began its mission, he reminded his charges, "Our God-given mission as doctors is to challenge all varieties of disease-causing microorganisms, to block all roads of intrusion into the human body, to annihilate all foreign matter resident in our bodies, and to devise the most expeditious treatment possible." And now they faced a new world, where their mission "is the complete opposite of these principles, and may cause us some anguish as doctors."[34]

But he beseeched them to forge on. They should focus on *"the dual thrill":* they were scientists probing for the truth by their research and discovery in the unknown, and they were soldiers who would build powerful military weapons against "the enemy."[35]

By 1975, adoption of the Biological Weapons Convention seemed to herald the end for development and use of these hideous and unpredictable weapons. If only that had been so; for now the devil hid not in details or omissions but hid quite literally—under cover.

The BWC was indeed a "gentleman's agreement" and so lacked any enforcement provisions, such as on-site inspections. That failure prevented any BWC party from knowing what the others were up to, and at least two of them were up to plenty. For all its twists, turns, and missteps, the story of the United

States leading the world to the BWC is a model for how things can go right. And for all that the BWC got right, the lack of enforceable openness and transparency led to the story of South Africa, which became a model of how things can go horribly wrong, especially when a country is consumed by fear that an adversary is up to worse. In this case, the major, often covert, adversary was the Soviet Union, and it certainly was up to no good.

For Reasons of National Security: Project Coast

As Dr. Schalk Van Rensburg told it, he saw great good coming out of the research he had been recruited to in 1984.[36] An animal scientist with a special interest in cancer research whose work was highly regarded, Van Rensburg was brought into a vanguard South African research project by a fellow scientist whose research and ethical qualities he deeply admired. Dr. Daan Goosen told him they would start a model contract research company. Their work in human reproductive research, including searching for more effective contraceptives, would be of great use to the pharmaceutical industry. Certainly the World Health Organization and international human rights organizations were gravely concerned with the global population explosion, especially in poor nations.

The man at the top, Dr. Wouter Basson, would become notorious for the projects he sought—and demanded—as well as for his charisma and arrogance. But at the time, that seemed offset by the genuine needs of a nation imperiled on every side and from within.

Picture Van Rensburg as he might appear on a Sunday morning talk show, describing the heady, early days as they anticipated great discoveries: We had fantastic scientists and a very high percentage of good trained people in this country. We had the most modern animal research technology.

> We had the reputations, everything was going and we thought we could create a fine facility that ... could be an asset to the country from an industrial point of view, it could earn a lot of foreign exchange and create a lot of jobs.... This against the background of failing funds for basic research work.... It was actually a world-wide trend, that government funds for basic research, particularly my field—which was the causes of cancer—these were being cut all over the world.[37]

In what would become known as "Project Coast," they believed they would get to do excellent work in well-equipped laboratories. They would not lack for funding, and they would be well paid, though not far above their colleagues in other institutions. For him and several others, it was never about personal gain.

But this is no talk show we're overhearing. It's the South African Truth and Reconciliation Commission, headed by Bishop Desmond Tutu, surely one of the most remarkable examples of national atonement in memory, and the testimony of Schalk Van Rensburg and a handful of other scientists stands out for its implications for the future of science in a time of war.

From the start there was another dimension to the work Van Rensburg and his coworkers would carry out, and it was covert. They would conduct secret projects for the military. But this was a time when every South African of every race and political persuasion felt deeply threatened, even if not in the same ways.

Despite a range of political beliefs among the scientists, "all were able to justify their activities in terms of the need to respond to the war that most white South Africans believed themselves to be engaged in," Van Rensburg testified. "It is impossible to fully understand their motivation without accepting that this understanding was prevalent amongst white South Africans."[38]

There were contrasting forces pulling at Van Rensburg from the start, and the covert work troubled him. But years earlier he had been forced from a long-held post by the right-wing Afrikaner Broederbond, so he thought he must be being recruited to an ethical endeavor. "I'm a pretty known liberal, blacklisted by the Broederbond, Anglican, lifelong Anglican," he recalled thinking. "There surely can't be anything very sinister if they ask a chap like myself."[39]

But he also had deeper, more personal concerns. One of his sons was a conscript in the army at the time. "People on patrol were definitely exposed to toxic substances, but they could never find out what, and it did cause occasional deaths."[40]

Basson had told him that chemical and biological weapons had gotten into the hands of anti-South African forces in Mozambique and Angola, where the Cuban army was known to be training insurgents with the support of the Soviet Union.

Behind it all was fear of the Soviet Union. For Americans, the internal struggle in South Africa may always be characterized by the conflict between the apartheid regime and the African National Congress. But for moderate South Africans, that was a less immediate issue.

Chandré Gould, a social scientist and human rights activist who would become an investigator for the Truth and Reconciliation Commission, debriefing the scientists who testified, told us the African National Congress "was a far lesser threat than the Soviet Union."[41]

But a military threat? "Absolutely. Namibia was a line of defense. Zimbabwe and Mozambique had eroded."

She says there were "certainly justifiable reasons for the Apartheid government and citizens to be worried about the military threat against South Africa." But American agents had similar fears. Years afterward as documents were declassified, "I looked at the CIA threat assessments in Angola, and they were certainly worried then."

And Van Rensburg had expertise on fungal toxins that was critical. He testified, "We knew the Russians were working on a new generation of chemical weapons based on fungal toxins ... and the one most promising product was Yellow Rain.... Minute amounts of these same toxins that were used for warfare sometimes occur in food and they can cause fatal human disease, so we were researching this intensively." But it is now believed by many scientists that Yellow Rain was bee feces—not a weapon at all—and Van Rensburg clearly should have known this at the time of his 1998 testimony.[42]

For Van Rensburg, the covert work, however troublesome, was necessary: "It is acceptable for any model army—in fact it is expected—to have a good technology base of covert warfare. If you don't know the dirty tricks you can't counter them, or you don't detect them, or you're outwitted."

So it began. But before Van Rensburg was fired, he would watch Project Coast's schemes deteriorate into a B-movie plot. He did not go along often enough and protested too much. And he quickly realized that the covert defensive work that was paying for the forward-looking research was actually 90 percent of all that was being done.

Among the projects put before him was to develop a contraceptive that would work only on black women, an idea he found so scientifically ludicrous he did not oppose it. He believed there was no way for a chemical to distinguish between human races. Apartheid leaders could wish for whatever they wanted and describe the product however they chose, but whatever he might develop would work on anyone or no one.

The proposal put to him, of course, was not promoted as racist. Basson told him that Jonas Savimbi, the West's guerrilla ally against Communists in Angola, relied heavily on his women fighters, considering them to be better than the men. But they were always getting pregnant, dangerously reducing his strength in the field. He wanted something to administer secretly that would prevent conception. Even if that were true, Van Rensburg reasoned, there would be no way to administer a contraceptive without a woman's knowledge.

Instead, he considered that anything he might succeed at could be openly offered—to anyone, anywhere. But that was just one measure of where Project Coast had gone. The labs turned out tons of the street drug ecstasy, ostensibly to control crowds by making them lethargic. Poisons were soaked into cigarettes and whisky for assassination purposes. From the inspired the project plunged to the pathetic—or maybe it had been pathetic all along. Looking back critically on the reams of testimony from the Truth and Reconciliation Commission, the entire endeavor seems a waste of enormous resources—scientific and financial—that accomplished little. And that was the good side of the outcome.

If Unit 731 demonstrates the horror that can result from putting too much faith in a group because of its medical expertise and presumption of moral superiority, South Africa is a far more recent example of how even good scientists can be lured, some unwittingly, by a genuine threat to national sovereignty and—every researcher's siren call—the promise of solid, permanent funding to carry out important work in well-equipped laboratories. It's a simple song, but, surprisingly often, it works.

Jeanne Guillemin wrote in 2007, "the capacity of scientists to set aside moral scruples is abundantly illustrated in the history of biological weapons in the last century, when thousands of microbiologists were employed in secret state programs that defied international norms and laws protecting civilians in war."[43] How about the Russians, then? Was it all world conquest in the name of Communism for them? Apparently not. For the most part, the prime motivator was the reason many players we have seen dive into massive biological weapons development: fear.

Beating the West at Its Own Game?

Ambassador Leonard reflected of the Soviets, "In the spring of 1970, as my memory goes, they simply reversed course and said, 'Now we're ready to negotiate as long as we don't have to do any verification.'" But Leonard believed, as he recalled the administration did, that the Americans had persuaded the Russians to end their program. "We knew they had a BW program and we thought we had persuaded them that they should do like us and dismantle it."[44]

But a rising young scientist intervened. According to Ken Alibek, Yuri Anatolievich Ovchinnikov, a young Soviet Academy of Sciences official, persuaded Soviet Premier Leonid Brezhnev in 1973 that the U.S.S.R. needed to explore "the military implications of the new gene splicing technology," and that exploration "led to the most ambitious Soviet arms program since the development of the hydrogen bomb."[45]

To develop a fully functional, well-funded, top-secret biological weapons program aimed at creating the deadliest strains of pathogens and the means to turn them into weapons of mass death, one formula seems to work every time. In this iteration, a double-agent spy may have fueled suspicions and fear about U.S. intentions.

Ambassador Leonard recalled a story worthy of John le Carré at his finest, a "super, super secret" revealed in a 1973 briefing given to only six or seven top administration officials that they were told never to reveal. Earlier in the 1970s, the FBI had a Russian double agent they would feed disinformation to be passed on for Soviet consumption.

"For some reason they decided that they would feed a story to him that the whole BW treaty was a sham ... and that we really were massively violating it," Leonard said. He still has no idea what the FBI agents' motive would have been, recalling, "we were all horrified, and the FBI of course was sternly reproached by the administration." Was damage done? Was this the cause or a contributor to the Soviets' disbelief in American intentions?

Although that information has since come out, Leonard says to his knowledge, no one has ever looked into the consequences of the bogus story. "It is possible that ... the people at the other end in Moscow realized the man was a double agent and dismissed the whole thing," he says. "But it seems rather more likely that this contributed to a feeling on their part," that just as Alibek has related, as far as the BWC went, "there was a total conviction that the American signature was a sham."[46]

Remember that this was four years after Nixon boldly renounced their development by the United States. But why should the Russians believe that the capitalist conspirators would keep their word? There was no way of knowing what other nations were really up to, before or after the Biological Weapons Convention was passed.

Ironically, the Soviet Union used the BWC as a cover for its massive offensive bioweapons program, founding Biopreparat, billed as a civilian drug company, just one year after signing it. Just like Project Coast a decade later, Biopreparat openly produced pharmaceuticals and covertly studied and built biological weapons.

At its peak, the Soviet Union had the world's largest bioweapons program, with somewhere between twenty-five thousand and thirty-two thousand scientists and support staff employed in a network of twenty to thirty military and civilian laboratories and research institutions that had spread throughout the republics, with scientists cooking up a nightmare list of deadly pathogens, including that worst of historical killers, smallpox.

Biopreparat was charged with the task of developing biological weapons genetically engineered to be resistant to antibiotics and vaccines. Soviet scientists primarily focused on developing the agents for tularemia, plague, anthrax, and glanders. Apart from working on agents that targeted humans, Soviet scientists also developed bioweapons that could be used against crops and animals. When sprayed over livestock and crops, these diseases could cause enormous economic damage, and—in nations where hunger is the norm—massive death from starvation.

Much of what is known about the secret Soviet program has come from high-level figures who defected to the United States, notably Ken Alibek,[47] Biopreparat's former deputy director. Through Alibek's eyes, we get a close up view of why the massive Soviet offensive bioweapons program started and how he got into it.

Alibek began graduate studies as a "cadet intern" in 1973 with the aim of becoming a military psychiatrist. Then, while taking a course in epidemiology, he was given an assignment that would radically change his life. He was asked to analyze an outbreak of tularemia, the often-deadly bacterial disease called rabbit fever, that had stricken Stalingrad during World War II. After examining the epidemiological data, he came to the surprising conclusion that the outbreak was actually caused by Soviet troops, who intentionally released tularemia in a desperate effort to halt the German advance—a conclusion that today is highly disputed.[48]

Suddenly Alibek found himself absorbed by bioweapons and their potential. He began to focus his studies as a military cadet on infectious diseases and epidemiology, and he caught the eye of Biopreparat recruiters. Offered the opportunity to continue this work, he accepted and headed down an unforeseen path to the highest levels of authority over this largest of all bioweapons programs.

Alibek was initially concerned about the ethical implications of his Biopreparat work. He and his fellow physician recruits struggled to reconcile the Hippocratic oath to "do no harm" with their new goal of producing highly lethal biological agents. Alibek reflected, "I still shuddered occasionally when I looked at the bacteria multiplying in our fermenters and considered that they could end the lives of millions of people."[49]

Alibek's love for the exciting scientific work he was doing, however, eventually won out over any ethical qualms that may have held him back as a doctor.

"The government I served perceived no contradiction between the oath every doctor takes to preserve life and our preparations for mass murder. For a long time, neither did I."[50] He said that "the idealistic young doctor ... who had agonized over the difference between saving lives and taking them, was gone."[51]

Then came the end. The fall of the U.S.S.R., so welcomed in the West, also opened a Pandora's box of loosely guarded biocontainment facilities falling into leaky ruin and left scientists with no means of support. There were concerns in the West that these desperate life scientists would be courted by a host of cash-laden foreign "shoppers." Fortunately, there is yet no credible evidence that Soviet bioweapons or expertise were sold. Where is the Soviet Union smallpox stock now? Presumably still only in its single Soviet era location.

As the past suggests, the United States may be sliding down the path it once condemned. Drawing U.S. scientists into secret research takes a page from Project Coast. And the growing paranoia over others' veracity might easily conform to the Soviet template. From day to the cover of darkness again.

But unlike the natural succession of days and nights, relentless yet at worst benign, we will see that what is brewing under cover as we enter the twenty-first century could permanently put out the light.

CHAPTER FOUR

Devils We've Known

While South African researchers, blinded by calls to patriotism, worked to develop "defensive" bioweapons, thousands of miles away another team of scientists zealously toiled to weaponize the world's deadliest bacterial toxin—to deploy against its own civilian population. They were recruits to a Japanese cult named Aum Shinrikyo—Supreme Truth—and how they moved from pacifist roots to follow their messianic leader into dreams of mass murder offers a parable of unquestioning obedience and the illogic of fanaticism. But that they worked for years hidden in plain view and that despite large-scale funding, intelligence, and dedication they utterly failed offers a far more important lesson: while many scientists might build a biological weapon of sorts, it takes more than run-of-the-mill scientific skills—no matter how ardent or twisted the faith—to build a biological weapon of mass death. Such a task demands the skills and knowledge only a few sophisticated offensive bioweapons programs had mustered.

A Holocaust to Save the World

The killer to be delivered was botulinum toxin, the cause of botulism, which has the ominous distinction of being the most poisonous substance known. One raindrop's worth of botulin spread on the tongues of a populace has the power to kill thousands. The acolytes of Aum Shinrikyo foresaw the enormous terrorist assault they planned as a means to hasten the Armageddon they believed was inevitable and out of which their leader would emerge to lead survivors into a peaceful future.

Deliver they did. Had they been successful, the botulin attack might stand as the worst terrorist assault in history. Instead, as far as it is known, no one died and no one was injured. As early as April 1990, the cult tried to release botulin from a vehicle driving around the Japanese parliament and other government buildings in central Tokyo. In early June 1993, members tried again as throngs gathered for the wedding of the crown prince. A vehicle equipped with a spray device was driven around the imperial palace.

Botulin was not the only murder weapon they worked on. That same month, the cult released anthrax spores from its midrise Tokyo office building, which housed its laboratory. Police and news reports told of foul smells, brown steam, stains on cars and sidewalks, and the death of pets. A bioweapons attack was not suspected, so no one looked for a culprit. More surprisingly, less than a year later in Matsumoto, the cult released a cloud of sarin vapor, a deadly nerve agent, at the quarters housing *all three judges* who were hearing a real estate dispute in which Aum

Shinrikyo was the defendant, leaving the jurists unscathed but killing seven people and sickening five hundred.[1] Still, no one from Aum Shinrikyo was charged with the crime. All that was about to change.

The cult's zealots now switched to a chemical weapon and finally gained international notoriety as terrorists. During rush hour on a Monday morning, the group released sarin into the Tokyo subway system. The attack left twelve commuters dead, nearly fifty seriously injured, and a thousand with lesser injuries.[2]

The March 20, 1995, sarin attack terrorized the country and shocked the world, but many more deaths and serious casualties would have resulted from conventional weapons costing virtually nothing. That would be tragically demonstrated just a month later. On April 19, Timothy McVeigh killed 168 children and adults in Oklahoma City with a homemade bomb made from nothing but fertilizer and heating fuel.

Shoko Asahara, founder of Aum Shinrikyo, and other leaders of the cult finally were sentenced to death for the subway attacks, but the sentences have yet to be carried out.[3] The cult is not only still operating but has been in the news in recent years after a splinter group following a new leader broke away from the main body—whose members still worship Asahara as the Messiah.[4]

Far from being a small band of violent political agitators or a shadowy al-Qaeda – like group hiding in caves, Aum Shinrikyo was a major international religious cult that claimed twenty thousand to forty thousand members and a net worth estimated by a leader at $1.5 billion. The money was collected through donations, tithing, sales of religious paraphernalia, and video and book sales, among other sources.[5]

A registered religious organization in Japan, Aum Shinrikyo had scientists and technicians carrying out deadly gas attacks and failed biowarfare attempts who were living openly, indistinguishable from their fellow citizens as they went about their work. Were the killers among them, then, religious fanatics who learned the tools of science? On the contrary. According to the Council on Foreign Relations, "The cult recruited bright young university graduates, particularly scientists, and put them to work developing biological and chemical weapons."[6]

Even if the $1.5 billion figure is exaggerated, the cult had huge sums to devote to its chemical and biological weapons program. Recruited scientists set up two labs built by Aum Shinrikyo engineers, stocking them with expensive equipment to experiment with, and they produced enough sarin to kill an estimated 4.2 million people, as well as phosgene and hydrogen cyanide, chemical stock of the infamous Zyklon B used in Nazi gas chambers.[7]

In addition to botulin, cult scientists experimented with anthrax, cholera, and Q fever, another barnyard bacteria – caused illness that is rarely fatal in humans.[8] In 1993, Asahara took a group of sixteen cult doctors and nurses to Zaire on a supposed medical mission. The actual purpose of the trip was to learn as much as possible about the deadly Ebola virus.[9]

No one knows what went wrong with the botulism plot, but some speculate that the cult's strain of the bacterium *Clostridium botulinum* produced little or no botulin, or that the toxin was destroyed when it was isolated from the bacterium. And the *Bacillus anthracis* they used to spread anthrax was a harmless vaccine strain—possibly the only one they could acquire.[10]

But there's more to Aum Shinrikyo's failure than a simple succession of mishaps. At every step in its strange progression from pacifism to terrorism, the cult offers critical lessons for the future, not only about how terrorists operate but, perhaps more important, about the limits of what even well-financed but inexpert scientists can do on their own.

Death Agents

Anthrax, plague, and tularemia derive from bacteria; smallpox, Ebola, Marburg, and Lassa are viruses; botulin and ricin are toxins. Here are microbes and poisons that instill fear at their very mention as we try to conjure their potential for havoc. However, in many ways they are as different from one another as biological agents are from conventional weapons. All but ricin share the distinction of composing the Centers for Disease Control and Prevention's list of Category A agents—the most deadly of the menagerie.

The CDC asked experts to rank the pathogens with highest potential as biological weapons into three categories. Category A agents threaten the most harm. We have added ricin to this group, because we believe it bears far closer watching than it receives. Bacteria and viruses, although far apart on the scale of life, do their damage to other living things simply by succeeding all too well at the first evolutionary requirement: thrive and multiply. These unfortunately do so at our expense. The bacterium *C. botulinum* makes the protein botulin that—probably only by accident—is a deadly neurotoxin in humans. Ricin is simply a component of castor oil plants with highly poisonous beans. They are flowering plants popular in gardens. As few as three beans can kill an infant.[11]

Different as they are, the world has a history with all these killers. Each member of the group has lethal traits that would make it attractive for biowarfare, and each has drawbacks that a would-be weapons builder would have to surmount. They now take the stage.

ANTHRAX

For Americans, anthrax remains the quintessential biological weapon, a terrifying complement to the 9/11 attacks. The four anthrax letters mailed less than a month later claimed five lives among eleven victims who developed the inhalational form of the illness, by far the most lethal. Eleven others developed cutaneous anthrax, or, infections through the skin.[12] Many of the survivors now live in failing health, another thirty-two thousand people had to undergo courses of antibiotics,[13] and the entire nation was terrorized—the defining goal of terrorism.

But for all of that, anthrax is a paradox. In its natural form it is a disease of cattle, sheep, and other grazing animals, and it almost never infects people in the United States. The few times it has done so, the victims typically were wool sorters, tannery workers, and others who handle unprocessed hides. In 2006, dancer Vado Diamonde became the exception that proves the rule, the first person in the United States since 1978 to contract inhalational anthrax from a natural source. He survived. Diamonde made drums from unprocessed animal skins, and that became the key element in a further case.

In September 2007, two family members from Danbury, Connecticut, one also a drum maker, were diagnosed with cutaneous anthrax from imported animal hides.[14] Both were successfully treated, and an expensive and lengthy cleanup of their property was completed.

It is perhaps surprising that two unrelated incidents, one in New York and the other in Danbury, occurred only a year apart, when the last reported case had been 1978. This may be because doctors now are alert to anthrax symptoms. If so, perhaps natural anthrax infections in the United States have always been more prevalent than believed.

The ubiquity of this microorganism in soils around ranches and feedlots hints at how harmless anthrax is to humans in developed countries—normally. However, the anthrax attack offers an important cautionary tale of humans' relations with natural microbes. With sophisticated laboratory manipulations by skilled scientists, even ultra rare diseases in nature that are not public health threats can become fearsome biological weapons.

B. anthracis lives in soil as spores that can survive for decades or even centuries. But if a grazing cow or sheep ingests some and is unlucky, the spores transform into reproducing cells and spread, killing the animal—which "to dust returns." Into that dust with the decaying animal goes the deadly bacterium, reverting to hardy spores. It is this long-term stability of its spores that helps make anthrax stand out among biological weapons agents, making it the first choice of nations with the will and the technical skills to develop it.[15]

And that brings us to the crucial issue for a builder of biological weapons: Among this lethal rogues' gallery, we need to know which would make the best recruits for weapons of mass death—that is, those that kill people but leave buildings untouched. Here is how anthrax measures up on a wish list for biowarfare agents. The ideal agent has the following characteristics: [16]

- Can be spread quickly over a huge populated area to strike a maximum number of victims, suggesting an aerosol or something highly contagious. Check: anthrax can be aerosolized, and if sprayed from an aircraft it could envelope large areas.

- Won't backfire on the attackers. Check again: anthrax is not contagious.
- Is stable enough to be stockpiled as a weapon, for years if need be. Another check: anthrax is that stable in its natural spore form.
- Kills its victims or makes them extremely ill. Check: a true terror on this score, aerosolized, inhalational anthrax kills nearly 90 percent of its victims who inhale enough spores *and* who do not get treatment.[17]
- Easy and cheap to make in large batches. Check on one score: easy to make, but difficult to weaponize as a powder fine enough to be deeply inhaled.
- Has the potential to create fear and panic. Already checked off the list: accomplished through the anthrax letters attack, though without causing widespread death.

By taking spores from the deadliest strains previously isolated from the soil, weaponizing them by secret processes to spread like smoke in the air we breathe and to penetrate the lungs deeply, the mass killer might emerge. We learned from the letter attacks aimed at U.S. lawmakers and others that anthrax that is not even weaponized can kill. And this is an important point to underscore here, at variance from some earlier speculation that found its way into news reports: the anthrax powder used in the attacks was fairly pure, but it was not weaponized to be delivered as an aerosol.[18]

On the bright side, we also learned that the infection can be cured even after symptoms appear, providing victims are given aggressive antibiotic therapy and good hospital care. However, sufferers may never fully recover. Years after the letter attacks, many victims still had not regained their health:

> "Some days I get up, and after an hour and a half I have to lie back down," says David R. Hose (case 20), 61, who was infected on his job at a State Department mail-handling facility.[19]
>
> Before he breathed in the microscopic spores, Hose says in an exhausted voice from his home in Winchester, Va., he was a healthy man who routinely put in 12-hour days handling heavy diplomatic mail pouches. Today, after the anthrax and a near-fatal bout of pneumonia, "I'm on three heart medications. I have asthma. I'm extremely weak."
>
> "These guys are also victims of terrorism," says Ramesh Patel, whose wife, Jyotsna Patel (case 13), 45, a New Jersey postal worker, survived inhalation anthrax and is still tormented by weakness, nightmares and crying bouts.
>
> Fatigue, says New Jersey postal worker Norma J. Wallace (case 11), 58, "is a given at this point. The shortness of breath still comes. I still have joint pain."

> By memorizing Bible verses and working through books of brain-teasers, Wallace says, she believes she has nearly overcome the memory problems that trouble the survivors.
>
> Still, "I have to consciously focus on what I'm doing, or I lose my train of thought."[20]

But remember that a weapon of mass death needs to be easy and cheap to make in large quantities. Deadly as it can be made to be, anthrax would present extraordinary problems for anyone trying to manufacture mass quantities, and that is a crucial qualifier. As far as we know, only the United States and the Soviet Union figured out how to carry out this process. Even Saddam Hussein, with his ambitious bioweapons program in the 1980s, likely had not yet successfully weaponized anthrax as a fine powder, although he did fill missiles with liquid anthrax.[21]

SMALLPOX

Until the end of the century, smallpox was perhaps the most feared disease on Earth, and no wonder.[22] It killed over two hundred million people during the twentieth century alone, more than a hundred years after a vaccine was developed. *Variola major,* the strain of virus that causes the most deadly form of smallpox, kills nearly one-third of its victims, and there is no treatment.

Hard as it is to imagine, until the end of the eighteenth century, most of our ancestors grew to adulthood with faces and torsos pitted with the deep scars of smallpox—unless they were milkmaids. No one understood why they had "milkmaid's skin" until Dr. Edward Jenner realized they had been infected by the mild disease cowpox, and that had inoculated them against its deadly relative. Jenner's cowpox-based vaccine virtually wiped out smallpox in the developed world. Elsewhere, it raged on.

At long last, an exhaustive, ten-year world effort eradicated smallpox from the planet thirty years ago, the last known case occurring in Somalia in 1977.[23] So ended a scourge that had killed and maimed millions for centuries and destroyed native populations lacking resistance, marking one of mankind's major public health achievements. But like all great achievements, the eradication of smallpox could have deadly consequences in an era of bioweapons. None of the world's young people have been vaccinated. They have no resistance to smallpox. Why does that matter?

When the Soviet Union collapsed, Colonel Kanatjan Alibekov emerged from its organizational rubble with stunning revelations about the extent of the nation's bioweapons program. Defecting to the United States and changing his name to Ken Alibek, the former deputy director of Biopreparat reported that the program had produced massive quantities of smallpox at its Vector facility, not simply a laboratory research supply.

Thus more than thirty years after its destruction in the wild, smallpox is not truly gone, it is only deeply hidden from view. The two known remaining stocks are at the U.S. Centers for Disease Control and Prevention and Russia's Vector. The World Health Organization, joined by many nations, is calling for the destruction of the two stocks as a permanent guarantee against the virus's being reintroduced accidentally or malevolently.

Developing nations, worried that they could do little to control a smallpox epidemic, argue for destruction, and for humanitarian and public health reasons many developed nations agree.

But there is a counter argument. What if a smallpox supply was stolen for weaponry twenty years ago as the Soviet Union collapsed? What if even now it were to escape from a U.S or Russian lab where it is being reproduced for defensive research? What if other nations never destroyed their stocks to begin with? Those are major reasons, some argue equally vehemently, that the existing stocks must be retained to develop drugs and new vaccines to protect against smallpox's return. However, having a viral supply does not guarantee having a new, safer vaccine at the ready. As Martin Furmanski points out, "The problem is that there is no useful animal model for smallpox," so efficacy of a new drug or vaccine would be difficult to determine in advance of an actual smallpox epidemic.[24]

Beyond question, reintroducing smallpox to the planet would be among the worst crimes ever committed against humanity. But is that awful scenario best countered by preserving or destroying the virus? That is a question we all must work to answer. Meanwhile, a group of experts convened by the *Journal of the American Medical Association* wrote of the threat from a smallpox bioweapon that "its potential for devastation today is far greater than at any previous time." Americans have not been routinely vaccinated for more than a generation, the experts said, noting ominously, "In a now highly susceptible, mobile population, smallpox would be able to spread widely and rapidly throughout this country and the world."[25]

In a developed nation with sufficient supplies of smallpox vaccine, an epidemic might be averted with early warning from the first few cases; however, in a developing country where there is likely little or no smallpox vaccine, the disease would spread quickly. That alone, however, should temper some officials' fears of that agent being spread by Islamic extremists. Furmanski says, "If anybody is vulnerable to a smallpox outbreak, it is the Islamic world, where populations are young, the older population with some residual immunity from vaccination prior to the 1980s is small, living conditions are crowded, vaccine stocks were absent or grossly inadequate, and health care facilities for isolation of active cases are scarce. So much for smallpox being attractive to terrorists from that camp."[26]

PLAGUE

From the epidemic-bearer that swept Europe and claimed a quarter of its population to the battlefields of China where it became endemic in villages that had never known it, plague has become established in the human population, as have the rats that generally harbor the fleas that often carry the bacterium *Yersinia pestis.*[27] It's a complex chain of host and hosted, as we saw in chapter 1.

Plague is also a bacterial disease, yet unlike anthrax, one form is highly contagious, though it is not the form we think of. We associate "plague" with "bubonic," but that form is the least lethal and least contagious, usually spreading from a rat flea to attack the lymphatic system of the victim, who breaks out in ugly blackened pustules, the eponymous *buboes.* Another form, septicemic, infiltrates the bloodstream. But it is pneumonic plague, the *Y. pestis* infection of the lungs, that is both extremely deadly and highly contagious—the bringer of mass death. Victims can catch pneumonic plague from a victim's cough or simply from having close contact. Luckily, after the Middle Ages, it also became rarer.

Plague is endemic in many areas of the United States, though seldom fatal, because, unlike the Chinese victims in Manchuria, sufferers today with access to clinics can get antibiotics to clear the infection and can be quickly isolated to prevent contagion. The incidence here is low. Since 1947, a total of 390 cases have been reported—about 60 a year—and only about 60 altogether died in that time span.

If that seems to render plague of no concern, the *JAMA* experts rank it among the most feared microbe in the rogues' gallery for several reasons. First, it is everywhere and therefore easy to obtain. Second, it has already proven itself in nature for aerosol dissemination—important for a truly devastating bioweapon—and indeed, that pneumonic form is far more lethal. Finally, the *JAMA* experts note, "The biological weapons programs of the United States and the Soviet Union developed techniques to aerosolize plague directly, eliminating dependence on the unpredictable flea vector."[28] Producing aerosols of viable plague is, fortunately, difficult, for the plague bacterium is fragile compared to anthrax spores.

Another saving grace for now is that *Y. pestis* can be killed with such common antibiotics as streptomycin and tetracycline. Thus, barring a massive initial infection in a bioweapons attack that overwhelms public health resources, plague is of little threat to developed nations with a readily available supply of antibiotics and quarantine and thoroughgoing sanitation systems. Plague is likely down the list of bioweapons choices.

RICIN AND BOTULINUM

The stories hit the news fairly often:

BBC NEWS WORLD EDITION
SEVENTH ARREST IN RICIN CASE
Wednesday, 8 January, 2003, 18:28 GMT

A seventh man has been arrested by anti-terrorist officers investigating the ricin find in London.... Six men—understood to be north Africans—were arrested on Sunday and security experts are trying to establish if they have links to al-Qaeda.[29]

And there's this:

FLORIDA MAN FACES BIOWEAPON CHARGE
FBI Says Accused Had Poison Ricin and Several Weapons
Friday, January 14, 2005 Posted: 6:48 AM EST (1148 GMT)

MIAMI, Florida (CNN)—An Ocala, Florida, man was arrested by the FBI after they found the biotoxin ricin in his possession in the home he shares with his mother.[30]

Most recently, the *Los Angeles Times* described a California man jailed in Las Vegas for ricin possession this way:

He had struggled with painkillers and alcohol. He was often depressed. In the late'90s, after a love affair ended, Roger Bergendorff turned to making the toxic powder ricin.

> "I felt this was a harmless outlet for my anger, a potential shield" against those who might harm him, Bergendorff said. "I could not have used it even if I wanted to, because I fear God's judgment."
>
> In February, Bergendorff was living in an Extended Stay America in Las Vegas when he was hospitalized with symptoms of congestive heart failure, court papers said....[A]uthorities recovered castor beans, whose processing waste is used to make ricin; a weapons cache; a copy of "The Anarchist Cookbook" with a page about ricin marked; and four "crude" grams of the toxin.[31]

Ricin and botulin: two biological toxins that appear in the news at opposite ends of the spectrum. A search of news archives turns up ricin often, but generally on an inside page. Botulin turns up rarely, but then typically on page 1. Within that gap lies a most important consideration for anyone undertaking risk assessments for different biological weapons. How lethal a potential agent is is obviously critical, but no more so than such factors as ease of use.

Castor beans, the source of ricin, are easy to get, as the castor oil plant is a popular flowering plants in gardens, so it has attracted the attention of terrorists and kooks alike. "Simple," of course, is a matter of perspective. Following available Internet recipes, it is simple for a chemist familiar with isolating desired compounds from the complexes they nest in, but not so simple for the average person. But with the will and the time, it can be done, as the news stories show.

In contrast, *C. botulinum* is an anaerobic bacterium, which means it can survive only where there is little or no oxygen. It will grow very poorly or die in open air. While some microbiologists are adept at culturing anaerobes, the work is not easy. Aum Shinrikyo chose botulin because it is the most potent toxin known—thirty times more so than ricin if swallowed. However, the strain of *C. botulinum* must actually be producing the botulin toxin protein, and that may have led to Aum Shinrikyo's failure.

However, different as they are, each can be made far more deadly as aerosols than if mixed in food. Both are proteins, meaning they are big molecules that are difficult to stabilize, yet they must be in order to be made into fine powders or sprays delivered via aircraft or missile.

So which is the more fearful potential weapon? Ricin is, by our reckoning, which brings us back to the reason we included it on our "most deadly" list. The relative ease of preparing and mixing ricin into food heavily weights the *likelihood* that it can be used *successfully* in an attack. In other words, risk assessment should not be based simply on the number of casualties and economic and other consequences a potential bioweapon might cause, but on the likelihood or probability of a successful attack as well.[32] In simple arithmetic terms:

$$\text{Risk} = \text{Likelihood} \times \text{Consequences}$$

"Likelihood" is the key word in assessing the likelihood-weighted risk—that is *the real risk*—of the panoply of microbes and biological toxins that in theory could be used against us. "Consequences" refers to the number of deaths and other harm. Unfortunately, the potential for the largest numbers of deaths and harm without regard to likelihood—that is, high consequence, low likelihood attacks—guided U.S. biodefense policy in the years after 9/11. However, that is now changing to the more rational, higher likelihood - weighted events. Even if they deliver fewer casualties, *high probability* attacks often represent a greater risk than *high consequence* attacks.

TULAREMIA

Rabbit fever is the outcome of bacterial infection by *Francisella tularensis,* and it is endemic throughout the United States.[33] However, fewer than two hundred cases of tularemia, the disease's formal name, are reported annually; only one or two victims die. But *F. tularensis* is one of the most infectious pathogenic bacteria known, and as few as ten bacteria inhaled or entering through skin breaks can cause sickness. The good news: it's not contagious.

Because it is so infectious, laboratory workers are especially vulnerable to getting tularemia. In one high-profile incident a few years ago, Boston University lab workers became ill from what they thought was a noninfectious strain. They were taking only minimal precautions in the lab. What might have been a minor incident deservedly made big news, because the university did not report it immediately. More troubling, Boston University was then beginning its campaign to build a BSL-*4* laboratory. What if a similar accident happened with a highly contagious bioweapons pathogen? When would residents find out?

EBOLA, MARBURG, AND LASSA

Rounding out the A-list are the terror viruses sensationalized on page and screen, the filoviruses Ebola and Marburg and the arenavirus Lassa, and they are far more famous than their fatal accomplishments warrant, thanks to *The Hot Zone* (1994) by Richard Preston, a years-long bestseller and the idea behind the major movie *Outbreak* (1995) that put their gruesome infections center stage.[34] The nonfiction narrative features vivid descriptions based on participant accounts of a 1980s outbreak of Ebola virus among lab monkeys that spread fear of an epidemic among CDC and Army officials. However, *The Hot Zone* has been criticized by scientists and biosecurity experts for both sensationalizing and far overstating the risk of a major human outbreak.[35] It is still unclear whether filoviruses can be an effective aerosolized bioweapon, but it seems unlikely.[36] The Soviets may have tested aerosolized Marburg virus,[37] but as a weapon it is likely significantly inferior to anthrax. Only a sophisticated bioweapons program could achieve weaponization of filoviruses—certainly not one put together by terrorists.

Contrary to their terrifying image, these viruses are not major public health threats even in Africa, where small outbreaks usually claim no more than a few dozen victims. Ebola and Marburg are not especially contagious because contact with a victim's fluids is necessary to become infected, and victims show symptoms so quickly that they are easily isolated, and they die quickly, before they can spread contagion over a wide area. All that is bad for victims but good for those even a slight distance from them.

What can be wrong with eyewitness accounts of an outbreak? *The Hot Zone* idealizes the virus as a human killing machine. It is anything but. According to Jens Kuhn, an authority on these microbes, filoviruses are not very contagious.[38] In addition, they are not very stable, making it unlikely that they would launch a major epidemic. Further, hemorrhages and liquefying organs are not typical symptoms. Perfect biological weapons? Not at all, he says. Kuhn also has spent time in Russia working with ex-Soviet bioweapons scientists, and from that experience he is convinced that former Biopreparat official Ken Alibek has overstated the bioweapons potential for the filoviruses in the Soviet program.[39]

But the filoviruses *are* horrific. They cause severe hemorrhagic fevers in which bleeding may occur from the mouth and other mucous membranes and orifices. They are extremely lethal, killing anywhere from half to 90 percent of their victims, depending on the strain.[40] But outbreaks of the filoviruses are localized on their continent of origin, Africa, where outbreaks kill dozens of people, not thousands. They are in no way international public health threats.

Apparently these two viruses do not pose biological weapon threats to the United States, either. Government biodefense researchers at the United States Army Medical Research Institute for Infectious Diseases (USAMRIID)—the organization made famous nationwide by *The Hot Zone*—say that most evidence suggests they are not very stable in aerosols, the major mechanism that would make them a bioweapons threat.[41]

Unlike Ebola, Marburg, and Lassa, many others on the Category A list have killed masses of humans in epidemics over the centuries. And they are not alone, for doctors and scientists keep warning us that the devils we know—AIDS, antibiotic-resistant pathogens, and emerging strains of the influenza virus—pose by far a greater threat to us as individuals and as a society.

Lessons

In their native form, the bacteria and viruses discussed as death agents are virtually no public health threat in developed nations. Even highly contagious agents like plague are only minor worries with good public health, sanitation, and plentiful antibiotics. This calls into question the mass deaths predicted by some biodefense spokespersons for bioweapons attacks using aerosols of available strains. Note that we are referring to these microbes as they exist in nature.

But if some of these agents can somehow be weaponized as aerosols they could cause a large number of casualties. Once weaponized, aerosols of bioweapons agents are deadly, because they can reach large numbers of victims and quickly enter the body. The biological toxins also present a threat to us, but as we will see, casualties are likely limited from an attack through poisoned food, the probable avenue of dissemination. What about poisoning our water? The less potent chemical agents are simply too diluted in water supplies to cause harm, and the potent protein agents like ricin and botulinum toxin would degrade before they arrived at the faucet.

In the end, weaponizing these agents for aerosol delivery is difficult and probably beyond the capabilities of most terrorists, as the wellfinanced Aum Shinrikyo cult vividly demonstrated. The sophistication to weaponize and deliver bioweapons lies in the well-equipped laboratories of nations, not the dens of terrorists.

But even such advanced nations could not develop bioweapons of proven military value in the era before the Biological Weapons Convention prohibited them. Martin Furmanski conducted an exhaustive, as yet unpublished, ten-year study of programs from that period, analyzing them from medical, technical, political, and military perspectives. He concluded that bioweapons programs in Japan, the United States, and the United Kingdom "were, overall, technical failures." Moreover, only a handful of agents were even claimed as successfully developed and standardized for production, and most of those—such as Q fever, Venezuelan equine encephalitis and *Brucella*—were not very lethal.

He added, "except perhaps for anthrax, most would likely fail if released in an urban environment."[42]

Those failures, of course, are reasons for optimism.

"Terror" by the Five-Pound Sack

What do you get when you cross terrorizing fear with cold calculation? That might be a working definition of terrorism. Consider its complement. Here is what you get when you pour out ominous possibilities straight, without stirring in the odds of their coming true.

On November 16, 1997, William S. Cohen, then defense secretary, appeared on the ABC network show "This Week," speaking to host Cokie Roberts in Washington, D.C. The world was then pressing Saddam Hussein to cooperate with UN inspection teams with the unwieldy acronym UNSCOM (United Nations Special Commission), which were searching for weapons of mass destruction, but he was not cooperating. Nevertheless, in 1995, UNSCOM *had* found evidence of bioweapons in Iraq, including anthrax and botulinum toxin at Al Hakam and camelpox and other viral agents at Al Daura.[43]

Ironically, the inspectors of UNMOVIC (the United Nations Monitoring, Verification, and Inspection Commission), which succeeded UNSCOM, did get full access to Iraqi facilities and far greater cooperation—but they did not uncover new evidence of bioweapons. The inspectors might have succeeded, given more time, if indeed there still were bioweapons to be found. They were forced to leave Iraq in 2003 as the United States was about to invade. However, as we now know so emphatically, no evidence of these or other "WMDs" ever was found after the invasion.

Reflecting on the question of whether Iraq still possessed bioweapons when America invaded, David Kay, the U.S. post-invasion weapons inspector said, "If there were large stockpiles, they had to be produced by people, they had to be produced in facilities, and they would have left some indelible signs. Where are those people? Where are those facilities? Where are the documents, the importation and the other records of such large production? They have not been found. And I think those are pretty compelling proof at this point ... that those don't exist."[44]

But let's get back to 1997. In order to alert Americans to *the real danger* posed by Iraq's biological and chemical weapons, Secretary Cohen held up a five-pound bag of sugar, as innocent a prop as could be found—and that was his point. Ordinary and everyday, this paper bag represented five pounds of anthrax.

Cohen said that amount could kill half the 550,000 inhabitants of the nation's capital.[45] "One breath and you are likely to face death in five days," he said.[46] But he wasn't finished yet. *New York Times* columnist Maureen Dowd wrote, "When he held up a vial, noting how deadly a single drop of VX nerve gas could be, Cokie Roberts begged him to put it away."[47]

Chemical weapons can kill large numbers of people—but only in large quantities delivered uniformly over a large populated area. *One raindrop's worth of botulin spread on the tongues of a neighborhood has the power to kill thousands.* But how would you spread such a raindrop so perfectly? Consider VX, the truly potent nerve agent, one vial of which Cohen claimed could kill nearly 300,000 people—about half the population of Washington, D.C.[48]—if we were attacked by Saddam Hussein. Indeed, VX is 170 times more potent but much less volatile than sarin,[49] the gas used by Aum Shinrikyo in the Tokyo subway attack.

How much VX would it take to kill half of Washington, D.C., and how would a would-be Saddam Hussein deliver it? Calculating from a report by Matthew Meselson,[50] the average lethal dosage of sarin over one square kilometer would be about 1,000 kilograms, or a metric ton. But VX is 170 times as toxic, so an attacker would need only 5.88 kilograms, or about 13 pounds, to get the same deadly effect with VX. Assuming the gas would kill roughly half of those exposed and that the population in Washington is distributed over about 159 square kilometers,[51] coverage would require nearly 159×5.88 kilograms. That's 935 kilograms—again, just under a ton. That amount could be made in a small, sophisticated, and very safe chemical plant that indeed could be tucked away almost anywhere.

Reason to be terrified? Not quite. Distributing such an amount of VX would be the stuff of science fiction, but still we will consider the only two ways to deliver a large amount of airborne weaponry of any kind: missiles and aircraft.

An attacker would need to arm and aim a large number of VX-packed missiles to blanket Washington in nerve gas. Let's estimate that a single warhead could distribute enough VX to kill within a radius of 450 feet—that's 636,000 square feet.[52] The 159 square kilometers comprising Washington figures out to some 1.7 billion square feet. So about 2,700 singlewarhead missiles would be needed to cover this large area.

Sophisticated missiles that pack more than one warhead have been in international inventories for many years, but each multiple of warheads in a single missile ups the technological and engineering know-how required of an attacker. Suppose, however, that an angry autocrat like Saddam Hussein builds such a fleet with ten warheads per missile. That still means 270 missiles inbound from overseas launch sites, each with its own "GPS home" clearly visible—perhaps even on Google Earth—and all of them tracked from liftoff by America's and other nations' space-, air-, and ground-based defense systems.

A bomber fleet scenario falls apart by the same logic. How about a 9/11type strategy? One midsize aircraft with a large, leak-proof storage tank carrying a ton of sprayable VX somehow evades detection. The design and construction of such an aircraft and spray system would require exceptional technical skills, but let's assume a rogue nation has them. That aircraft would need to make perhaps 100 evenly-spaced passes over Washington to blanket the city, just as crop dusters stitch their way back and forth over agricultural fields every summer. How long would it take for it to be shot down? Given that the target is Washington in the post-9/11 era, it would be shot down soon after beginning to spray VX. Whatever damage it would cause would be from exploding fuel and flying debris that could include VX spread over a small area. It would do less damage than if it had dropped a conventional bomb.

The richest country on Earth could not launch a massive VX attack in secret—or anything approximating it. How would Saddam Hussein have carried off that feat when he couldn't have manufactured WMDs or assembled such an armada without the world's knowing about it? How would a group of terrorists operating without an oil-rich nation's resources? This is one way of weighing possibilities, by looking at their *im* possibility.

Support for our argument comes from a particularly knowledgeable source. Colonel James Ketchum, MD, whom we encountered in our discussion of incapacitating agents, included an unsent letter to the *New York Times* in the text of his 2006 book *Chemical Warfare,* writing that he thought the letter had no chance of being published:

> I feel the [Bush] administration has been negligent in failing to explain the limitations of chemical weapons. As a physician who spent the better part of ten years conducting medical research with chemical agents during the'60s, I have followed closely the portrayal of nerve agents (such as VX) by government and media representatives. It seems to me that fear of death through exposure to these substances has become irrational, ignoring scientific reality.... The amount of VX in a warhead will normally be lethal (assuming a full minute of unprotected inhalation and optimal wind conditions) out to a radius of 100–150 yards....
>
> Why have there been no terrorist attacks with nerve agents in the many months since 9–11? The answer seems obvious: *chemical weapons are not particularly effective.* At best, they cause deaths in a circumscribed area where there is no protection and no escape. Two hundred kilograms of conventional high explosive, the amount deliv-

> ered by a SCUD missile, can cause more deaths than the same amount of "nerve gas." Any statement that such agents will cause tens of thousand of casualties is gross hyperbole. What is worse, it unnecessarily fans the flames of panic.[53]

Using the numbers in the Ketchum letter, we calculate it would take a few thousand warheads [54] delivered at a precise distance apart to kill 300,000 people. No terrorist group and few countries could mount—nor would they dream of mounting—such an attack. Conventional bombs would do the trick much more easily, with far less risk to the attackers.

The Usual Case

Let's consider another forecasting method, applicable to situations where we can't absolutely rule out scenarios: weighing the *relative* likelihood of different bioattack scenarios. Here is how it works. First, consider a lowtech ricin attack in which large numbers of restaurant salad bars are simultaneously laced with ricin, resulting in 1,500 fatalities. In fact, this would be a radically scaled up version of the poisoning of salad bars with salmonella in The Dalles, Oregon, area in 1984, which left 750 people ill but caused no deaths.[55] A second attack to be considered will be larger scale, a high-tech assault of the sort used by the government in disasterplanning calculations. For this, a kilogram of pure weaponized anthrax will be spread in the air under optimum conditions, resulting in 15,000 fatalities.

What we ultimately want to know—and the purpose of this exercise—is how to reasonably weigh a possible ricin attack against a possible anthrax attack. The first critical question is, what is the probability we will suffer a ricin attack in any one year? We have no concrete idea, *but it is far, far greater than that for a weaponized anthrax attack.* Ricin can be isolated from castor beans by someone with the skills of, say, a home beer brewer. Recipes and complete instructions for obtaining ricin from castor beans can be found on the Internet and terrorists know it, so carrying this off would be no problem for such zealots, *if* they use an Internet recipe that actually works. Let's just guess that there is a 10 percent chance of a ricin attack from this moment through the next year—that is, a probability of 0.1 per year. Looked at another way, using a college-level probability calculation, 6.6 years would need to pass to have a 50–50 chance of at least one ricin attack.

Now let's figure into the calculations that it is ten times easier to carry out the attack with ricin than with anthrax. Assigning such a relativedifficulty factor to the anthrax attack is reasonable. Weaponizing this bacterium for a massive attack involves government-classified treatment of its spores to make them float in the air like smoke, so that they will be drawn as deeply as possible into the lungs, a requirement to develop inhalational anthrax. From the lungs, the spores are transported into surrounding lymphatic tissue, where they multiply and spread to other organs. Even if you know the recipe for weaponizing anthrax, preparing a kilogram in such form without killing yourself or your neighbors requires considerable expertise. And working in a biosafety lab would be a good idea as well—a lab at least as high tech and sealed as, say, a small submarine, which would be very difficult for a terrorist to assemble from scratch in his apartment or cave. Alternatively, you could just vaccinate yourself and take lots of antibiotics, which may protect you but certainly would not shield your neighbors from aerosolized anthrax.

Admittedly, we're doing a lot of guesstimating and hand waving in picking ten as the "difficulty" factor in using anthrax versus ricin, but we'll show that there are valuable lessons for the public welfare lurking in such imprecise calculations. What that factor often means is that in any one year the chance of the anthrax attack is one in one hundred.

Let's multiply those numbers to get a weighted or real threat using the formula Risk=Likelihood×Consequences, and to keep the calculations simple without minimizing the seriousness of the subject, we will consider fatalities the only consequence of either attack.

For the ricin attack: Risk=0.1×1,500=150 likely fatalities

For the anthrax attack: Risk=0.01×15,000=150 likely fatalities

Conclusion: The real threat posed by the ricin attack is equal to that of an anthrax attack, so it should receive as much attention and resources in biodefense planning. Supporting our argument, a recent FBI compilation of terrorist threats between 2002 and 2005 concludes, "Ricin and the bacterial agent anthrax are emerging as the most prevalent agents involved in WMD investigations."[56]

However, the government has taken the simplistic position that we must put most of our effort and dollars into preparing for catastrophic events like a big anthrax attack, or multiple attacks in rapid succession resulting in casualties far greater than 15,000. But our analysis, crude as it is, shows why equal attention should be paid to the less dramatic and less publicized low-tech threats as well. It also illustrates the fact that risk assessment is one of the most complex tasks governments must carry out as they decide where to put limited resources.

Here are some of the factors that a complete risk assessment considers:

- What is the nature of the threat? This includes looking at critical factors such as whether a resulting disease is contagious.

- How long does the agent linger in the environment? How expensive is decontamination and how long does it take? This would factor in the long time and huge effort it took to decontaminate post offices after the anthrax letter attacks.
- What would be the toll of sickness and death? For those sickened, how long are they ill and how expensive is it to treat them? What is the probable cost of prevention or cure?
- How likely is such an attack by small groups of terrorists or other disaffected individuals? How about state-sponsored or otherwise wellbankrolled terrorists? A developing enemy state? A developed enemy nation?

So much for bioweapons. We must also balance our biosecurity spending and actions between the ever-present threats to public health—AIDS, tuberculosis, malaria, antibiotic-resistant infections, emerging flu strains—and biodefense. We still need to answer questions at least as critical to our national future as the bioattack horror stories that might play out. The questions are far more complex because they entail all the natural diseases we know and the devils we don't know, such as a new pandemic flu. These are important and gripping issues to consider later. They are questions for arms control and defense experts, for epidemiologists and public health experts.

The clear lessons so far concern what terrorists can and cannot do and the priority that should be given to natural infectious diseases.

Without the purposeful—or unwitting—help of a nation, terrorists have little access to bioweapons and little skill in their production, safe handling, and use. Contrast this with the availability and ease of use of nitrate-rich fertilizer for a Timothy McVeigh, or stolen explosives that can be turned into the improvised explosive devices that are Iraqi militants' weapons of choice and the most frequent cause of death and casualty to U.S. troops.

Whatever large-scale threat we face from bioweapons, it is a threat from nations. That real threat is being worsened by our own government's paranoia. Paranoia breeds. Before considering the ramifications of that truism, carry this parable forward as you contemplate the direction from which very real terrorist threats are likely to emerge:

Imagine you are a terrorist bent on destroying the World Trade Center and the Pentagon, icons of American power and influence. So you painstakingly reverse-engineer a Boeing 767 aircraft, calculating every screw and slot to micrometer tolerances. Then you and your team build four exact replicas, which you learn to fly. Taking off from a secret, meticulously constructed remote airstrip, you evade radar detection and succeed in crashing three of the planes into their targets.

Nonsense, as we know all too well. Success is born of simplicity. It is a universal strategem of terrorists—in fact, of guerrillas and other fighters in "asymmetric warfare"—to *take* rather than to *make,* or, if they must, to make weapons by jury-rigging available agents rather than researching and developing sophisticated new ones. A headline-grabbing example is the use of chlorine canisters coupled with explosives by Iraqi insurgents.[57] The canister attacks killed and injured several dozen, but it is unclear how many of the casualties were felled by the explosion rather than the chlorine. These kinds of attacks have not occurred recently, and perhaps they have been abandoned because insurgents realized they were ineffective and imposed an added risk on them.

Yet such a bizarre scenario is strangely analogous to the government's expectations for bioterrorists. Implicit in U.S. biodefense strategy is the presumption that terrorists living in caves or third-world slums are at this moment conceiving and producing biological weapons and delivery systems so sophisticated that our biodefense scientists must work to build them first—*in order to develop countermeasures against them.* They are creating risks in order to counter them.

CHAPTER FIVE

Paranoia Begets Permissiveness

> Of course the people don't want war ...[but] ... all you have to do is tell them they are being attacked, and denounce the pacifists for lack of patriotism, and exposing the country to greater danger.
>
> HERMANN GOERING, NAZI REICHSMARSCHALL

Fueling Paranoia

The U.S. biodefense program is spiraling down a dangerous path for two overarching reasons that make strange relations indeed: paranoia and permissiveness. Each taken alone can lead to dire consequences. Taken together, the end can be far worse. When one hand of public policy works against the other, unfortunate scientists have been caught in the middle. The combination led to personal humiliation and imprisonment for one and a great deal of suffering for others, when the free behavior permissively encouraged in academia on one hand ran afoul of stark paranoia in government on the other. Other scientists have had their reputations destroyed.

Paranoia is a keystone in our government's political policy of instilling fear to maintain a strong image in the war on terror. This policy has created an overblown fear of a massive terrorist attack with biological or chemical weapons.

To counter that exaggerated fear, billions in biodefense funding have been offered up—with no end in sight—to entice scientists to develop medical countermeasures such as antibiotics, antivirals, vaccines, and antidotes aimed specifically at bioweapons to protect us from a massive biological weapons attack that may never come.

BioShield 2004 was the first act passed by Congress to provide billions in funding directed to countermeasures for bioweapons agents. Unfortunately, BioShield 2004's rules and related federal agency strategy [1] almost prevent development of countermeasures that would have a market for natural infectious diseases, and those are bigger concerns for most of us. With little control over how this money will be spent or even oversight of actual spending, we are creating a permissive atmosphere for the most unthinking scientists to dream up research with dangerous bioweapons agents in order to win funding. All told, the United States has spent several billion since 2001 on this monumental, wrongheaded effort.[2]

Secretary Cohen's "sugar-anthrax" scare aimed to impress on Americans the importance of biodefense efforts. But if such fears are bad for our national psyche, the paranoia they fuel has far more serious repercussions internationally. Paranoia leads to secrecy, and secrecy tends to fuel paranoia among nations just when we need cooperation.

Take a look into the dynamics that drove the cold war, this time from a Russian perspective. Ken Alibek, the high-ranking defector from the collapsed Soviet Union's massive bioweapons program, told of his introduction by the KGB to the sinister world that would become his life. During an initial KGB interview, Alibek assured the officer that he knew he wasn't getting into "normal" work. The officer pushed deeper. "I have to inform you there exists an international treaty on biological warfare, which the Soviet Union has signed," he said, referring to the Biological Weapons Convention.

"According to that treaty no one is allowed to make biological weapons," a tipoff, if Alibek needed it, that he would be moving into that very work. "But the United States signed it too, and we believe the Americans are lying."

Alibek assured him that he believed the Americans were lying. He and his peers had been taught as schoolchildren that the capitalist world was "united in the single aim of destroying the Soviet Union." On that paranoid note, his career began.[3]

Fear drove South Africa. The white public feared a Communist takeover, feared the black majority, and feared their neighbors' aggression. Minister of Defense Magnus Malan noted that the "onslaught" was "communist-inspired, communist-planned and communist-supported," and he asserted, "The security of the

Republic of South Africa must be maintained by every possible means at our disposal."[4] That, as we know, meant Project Coast, among other means at the nation's disposal.

And President Bush would say in 2002, "Bioterrorism is a real threat to our country," adding, "Terrorist groups seek biological weapons." In the run-up to the invasion of Iraq, he stated, "we know some rogue states already have them."[5]

How serious a threat were such weapons? Here he one-upped Cohen, asserting, "Armed with a single vial of a biological agent, small groups of fanatics or failing states, could gain the power to threaten great nations, threaten the world peace."[6]

From Soviet Russia to apartheid South Africa to twenty-first-century America, almost nothing has changed in the message, its patriotic fervor, or the strident tone of its delivery. And there was some truth in all of those statements. That's why they were—and remain—so effective in instilling the kind of fear that leads to irrational responses to real threats.

Even the language of simple nouns and phrases like "weapons" and "mass destruction" can hide worlds of distinctions that must be made between entirely different threats that might come from entirely different quarters, delivered through entirely different means.

One biosecurity expert points out that the phrase "weapons of mass destruction" makes sense in a treaty context, when the intention is to ban the broadest range of weapons that might fit the definition. But, in his scholarly study examining the use of the phrase, W. Seth Carus points out that the threat that each of these very different weapons poses is being masked by the common term.[7]

And that turns the phrase "weapons of mass destruction" into a weapon a demagogue would warm to—a club for control through fear.

Those knowledgeable about WMDs break them into four categories: chemical, biological, radiological, and nuclear, usually referred to by their initials, CBRN. Only those in the nuclear category are truly weapons of mass destruction, some capable of leveling most of a large city outright or through subsequent fires and, most destructively of all, rendering the remains uninhabitable for decades.

In contrast, radiological weapons—so-called "dirty bombs"—are conventional high explosives mixed with radioactive materials that kill and sicken inhabitants and make an area uninhabitable, but they cause no more destruction than high explosives alone. Some experts call them "weapons of mass disruption."[8] The remaining two? Chemical and biological weapons cause no "destruction," acting entirely against living things.[9] More important, the most likely threat we face from them is through small-scale, concentrated acts that may terrorize a population but will not cause large-scale death—witness the sarin subway attack and the anthrax letters.

Let's focus on "bioterrorism," a word that has evoked unreasonable fears since the anthrax letter attacks. Biological weapons are a fearful prospect, as we have asserted from the beginning. For years experts in biosecurity have been fighting to eliminate the weaknesses in international bioweapons controls, which are legion. It's the linking of those weapons to terrorist development and manufacture that we decry. Terrorism and bioweapons are indeed a deadly combination—they are the powerful fuel for paranoia.

Even as many experts criticize current U.S. policy and grow ever-more skeptical of future plans, most in this small science policy community note the irony in their position: for years they struggled to get officials to take the threat of biological weapons seriously; now that they have, those very officials are looking in the wrong direction.

Milton Leitenberg of the Center for International and Security Studies at the University of Maryland worries that "the U.S. government has been using most of its money to prepare for the wrong contingency," dumping at least $50 billion since 2002 to counter a threat that is "systematically and deliberately exaggerated." Consider that former Senate Majority Leader Bill Frist described bioterrorism as "the greatest existential threat we have in the world today."[10]

"But how could he justify such a claim?" Leitenburg asks. "Is bioterrorism a greater existential threat than global climate change, global poverty levels, wars and conflicts, nuclear proliferation, ocean-quality deterioration, deforestation, desertification, depletion of freshwater aquifers or the balancing of population growth and food production?"[11] He points out that mundane scourges like AIDS, tuberculosis, malaria, measles, and cholera kill more than 11 million people each year.

Jack Melling, the former chief scientist for Britain's bioweapons defense program, has a sharp, tongue-in-cheek comment for our government-whipped fears. "For me ... concern about exposure to bioterrorist attack comes below 'health of the neighbour's cat' on my list of worries!"[12]

For terrorists to launch a sizable attack, they would need to acquire, culture, weaponize, and finally deliver deadly strains of bioweapons organisms against a large population, well outside their capabilities and exposing them to new, unknown dangers compared to those of the conventional weapons they use all too well. Our efforts have not gone unnoticed, however.

Ayman al-Zawahiri, al-Qaeda's second in command and a practicing physician, noted of biological weapons, "Despite their extreme danger, we only became aware of them when the enemy drew our attention to them by repeatedly expressing concern that they can be produced simply."[13] And even if the last part isn't true, who can blame him for believing it when the United States Congress spends billions of dollars to counter bioterrorism?

In one case, support from private foundations and nongovernmentorganizationsinadvertentlycontributed to our fears by funding a theater of bioterror enactment exercises intended to train government workers in response and communications—always vital goals—*and* to assess damage potentials from a real bioterror event. The last is where the theater came in, along with the unfortunate consequences. Titles often echoed a Hollywood fright list: Atlantic Storm, Dark Winter, and the ominously charming Sooner Spring. The scenarios invariably proved precisely what their creators set out to show. Let's look at two and then try to separate the wheat from the chaff.

The benignly named TOPOFF (Top Officials), involving high-ranking U.S. government officials, was carried out in Denver, Colorado. The exercise tested emergency response networks in the face of a massive plague outbreak. Curtain up.

> Day 1 (May 20). "The Colorado Department of Public Health and Environment receives information that increasing numbers of persons began to seek medical attention at Denver area hospitals for cough and fever during the evening of May 19.... By early in the afternoon of May 20, 500 persons with these symptoms have received medical care; 25 of the 500 have died."
>
> Day 2. The "Virtual News Network," created as part of the exercise, "reports that a 'national crash effort' is under way that aims to move large quantities of antibiotics to the region as the CDC

> brings in its 'national stockpile,' but the quantity of available antibiotics is uncertain.... A few hours later, a VNN story reports that hospitals are running out of antibiotics.... By the end of the day, 1,871 plague cases have occurred in persons throughout the United States, London, and Tokyo. Of these, 389 persons have died."
>
> Day 3. "Hospitals are understaffed and have insufficient antibiotics, ventilators, and beds to meet demand. They cannot manage the influx of sick patients into the hospitals.... By noon, there are reports of 3,060 U.S and international patients with pneumonic plague, 795 of whom have died."
>
> Day 4: "There are conflicting reports regarding the number of sick persons and dead persons. Some reports show an estimated 3,700 cases of pneumonic plague with 950 deaths. Others are reporting 14,000 cases and more than 2,000 deaths.... The TOPOFF exercise is terminated."[14]

An earlier exercise, Dark Winter,[15] took place in June 2001, before 9/11 and the anthrax letters. It, too, involved a feared and highly contagious agent: smallpox. In the initial attack, "3,000 people were infected with the smallpox virus during 3 simultaneous attacks in 3 separate shopping malls in Oklahoma city, Philadelphia, and Atlanta." At the end of the exercise, 16,000 people were infected, with one-third of those dying. The exercise projected 1 million dead only a month later, in the worst-case scenario.

Among the unfortunate but important lessons of these exercises is the woeful unpreparedness of the United States for a deadly *natural* epidemic, let alone a massive bioweapons attack.[16] However, they tell us nothing about the likelihood of the bioweapons attacks depicted in their scenarios, nor the likelihood that such attacks might play out so dramatically. More seriously, hidden behind the scenario of TOPOFF is the assumption that terrorists could not only successfully aerosolize and weaponize the plague bacterium into its deadliest and most communicable inhalational form, but that they could deliver it covertly and simultaneously in world capitals. The assumption behind Dark Winter was that terrorists could lay their hands on smallpox, still an unlikely possibility. Think back on the 9/11 parable of the hypothetical jet aircraft builders versus the hijackers that were.

Well, then, what's the harm? What's wrong with basing exercises on worst-case scenarios even if they are highly unlikely? First, that gives terrorists free of charge what they work with such villainous zeal to bring about: public fear. And the unhealthy fear such manipulations provoke is unwarranted.

A major result of such overblown fear is what it does to our sense of well-being and even our mental health. Bill Durodie, director of the International Centre for Security Analysis at King's College, London, notes that "fear of infection could pose a greater strain on social resources than the pathogens themselves."[17]

This was graphically demonstrated after the Aum Shinrikyo sarin attack on the Tokyo subway, when four thousand healthy, frightened people showed up at local hospitals and were "classified as psychogenic patients without real symptoms."[18]

Even if the future were to bring a bioterror attack to our soil with as many as a few thousand casualties, our individual risks would remain close to zero. But we would not see it that way. We might huddle indoors and begin to duct-tape our windows in a desperate attempt to protect ourselves. The fear of bioweapons instilled in us by incessant warnings will blow our risk perception entirely out of perspective. Add to this our lack of understanding of what bioweapons can do, and a major run on hospital emergency wards hundreds of miles away from an attack might strain resources for no reason.

Fear might be very bad for us, but it is absolutely essential for a government that has built biodefense into a more than $50 billion colossus,[19] and likely has succeeded only in putting the country at more risk.

In the Darkness of Night

TOPOFF and Dark Winter were meant to be public, to alert officials to the dangers of bioweapons and our lack of preparation for an attack. However, prior to the 2001 anthrax letters, at least three secret projects were carried out by the U.S. government. Some believe they violated the Biological Weapons Convention and may be why they were kept secret. The three certainly violated another agreement among the nations that are parties to the BWC, an agreement which requires them to report their annual biodefense activities in order to increase transparency and to build confidence in everyone's peaceful intentions.

A *New York Times* investigative piece in September 2001 revealed these three suspect U.S. projects, their code names right out of a spy novel: Jefferson, Bacchus, and Clear Vision. And that story prompted the government to come clean on the projects.

Here is how the three are described in a 2001 British American Security Information Council (BASIC) report.

> (a) The Jefferson Project: The U.S. government learned that Russia had developed a genetically modified strain of anthrax in 1997. The government applied to Russia to obtain some of the strain in order to test its existing anthrax vaccines against it [but was refused].... The U.S. government then planned to develop its own genetically modified anthrax strain against which to test its existing vaccines.... The West Jefferson, Ohio, laboratory of the Battelle Memorial Institute was selected to create the genetically altered anthrax....[Spokesperson] Victoria Clarke said the administration had not yet produced the strain and that it does not plan to begin work until interagency consultations, legal reviews, and congressional briefings are concluded [and] added that

> the reviews completed so far indicate that the work would be BTWC [Biological and Toxin Weapons Convention] compliant.
>
> (b) Project Bacchus: The U.S.... has built a biological-agent production facility in Nevada using commercially available parts, reportedly to demonstrate how easy it would be for others to construct such a plant. The project also apparently aimed to assess whether small production facilities produce 'signatures' that could be used for identification purposes....[A U.S. spokesperson] said the Nevada plant produced only simulated biological agents, which are benign.
>
> (c) Project Clear Vision: In this programme the CIA reportedly built and tested a 'mock' biological bomb—a copy of a Soviet-designed biological bomb—to see how well it dispersed agents. The CIA feared that the bomb was for sale on the international market and decided to build its own model after efforts to purchase the original bomb failed. Intelligence officials told the *New York Times* that the 'mock' bomb did not have a fuse or other weapons-related parts that would make it operational.[20]

Why did the United States undertake these projects to begin with? We suspect the shock of discovering not merely its existence but the full extent of the Soviet offensive bioweapons program put a scare into our defense community. This is certainly understandable, but unfortunately it probably contributed to the rush of paranoia related to bioweapons after the 2001 anthrax letter attacks.

All three projects were carried out during the Clinton administration. Were the president and his top security advisors in favor of such controversial endeavors? Perhaps, or perhaps they were not aware of them. As events unfolded, Defense spokespersons discussed the issue only after the *Times* story appeared.

Spinning Out of Control

The United States is building a biodefense empire that is putting us at greater risk than we face from an attack from terrorists or foreign powers in the foreseeable future. At the same time, a swelling cadre of academic scientists is beating a path to research Earth's deadliest microbes, and toxins, further increasing that risk.

How did we get into such a mess? The answer invariably comes back to those strange bedfellows, paranoia and permissiveness. Paranoia brought us to an overblown fear of a massive terrorist attack with biological or chemical weapons. Permissiveness emerged in the torrent of billions of dollars in biodefense funding with little control over how the funding would be spent.

We are now going to descend into the shadows of government secrecy to bring to light three dangers, markers for our government's slide from what was only recently a moral high ground into a quagmire produced by suspicious and possibly treaty-violating activities: proliferating high biocontainment laboratories designed to work on dangerous pathogens, draconian biodefense laws, and overzealous enforcement practices. The three are inextricably linked in dragging us down into ever-greater risk in the name of improved security.

Threats that Aren't

How can you spend $50 billion and end up decreasing American and world safety? Start with the BioThreat Characterization Center. It is meant to be the definitive place to plan defenses against an unexpected attack, a place to develop an understanding of what is threatening us and find ways to counter those threats. However, some activities originally planned for the new center look remarkably like those that would mark our building a new *offensive* bioweapons program, and that is the major risk. The activities include studies of aerosol dynamics, aerosol animal-model development, novel delivery of an agent, novel packaging, genetic engineering, and environmental stability.[21] They might encourage unfriendly nations to begin a biological arms race in the belief that they are joining one already engaged.

The mission of the BTCC and the larger, enveloping agency in which it is nested involve assessing risk and conducting "studies and laboratory experiments" to test vulnerabilities, impacts and decontamination technologies, and vaccine development.[22]

What sort of experiments? *Creating new types of bioweapons and finding how to weaponize and deliver them.* If that froze your attention, it did the same to numbers of U.S. bioweapons control experts and diplomats, as well as our allies. But, the government claimed, it's all right for us to do this because we are only trying to create deadly new germs and weaponize them in order to learn how to defend against them.

A senior foreign diplomat representing one of our staunchest allies declared that the BioThreat Characterization Center denoted an *offensive program,* save for actual production of large amounts of bioweapons.[23]

Offensive-looking activities planned under BTCC auspices include studying aerosol dynamics to determine how bioweapons spread in the air, and researching how they might be packaged and delivered and kept stable in munitions and in the environment. Genetically engineering existing microbes, investigating immune system regulation and response, and other openly stated activities speak to a new generation of bioweapons designed to disable our body's natural defenses against disease.

Over the years the diplomatic and intelligence communities have developed "signatures" for governments' biological programs that might be telltale markers for offensive bioweapons endeavors. Like picking out the trail sign of a dangerous predator from among those of harmless wildlife, spotting these signatures would put the world on alert. Examples of such signatures include: military involvement in a biological development program, adopting a policy of secrecy, carrying out genetic engineering in bioweapons agents, conducting aerosol studies, and working with largeanimal experimentation facilities.

The BioThreat Characterization Center plans to carry out every one of these activities, so displays every signature.

The warnings cited in chapter 1 by Ambassador James Leonard, U.S. delegation head at the BWC negotiations, Dr. Richard Spertzel, former deputy director of USAMRIID, and Maryland scholar Milton Leitenberg probe even more deeply into disturbing issues posed by American biodefense activities. In a guest commentary in the journal *Politics and the Life Sciences,* they noted that, "Taken together, many of the activities ... may constitute development [of bioweapons] in the guise of threat assessment, and they certainly will be interpreted that way," and they note of bioweapons, "Development is prohibited by the Biological Weapons Convention."[24]

Also potentially suspicious to foreign eyes is the directive that intelligence agencies study the types of genetically engineered "bugs" terrorists could be working on to mount an attack. "Surely, the 'intelligence community' is the least appropriate place in the U.S. government to 'carry out' such work—and the most likely to lack adequate oversight," they write.

And they ask a series of revealing questions that are paraphrased here:

Does the program as designed reflect the real level of threat from bioterrorists? They point out that recently declassified U.S. intelligence documents show that terrorist groups like al-Qaeda have no ability to work with even classical BW agents, let alone to genetically engineer weapons.

Will the work be published in peer reviewed journals or kept secret? The administration wouldn't approve a verification protocol for the Biological Weapons Convention in 2001, because that would have required inspections of its far more modest program, the three experts note diplomatically. But presumably if we create a monster microbe and learn to weaponize it, we will not be telling the world that we did it and by what means we did it.

If we move in secret, what will be the reaction of other countries—including Russia and China? Long after the collapse of the Soviet Union, the United States and others are still working to secure seed cultures of Russian bioweapons agents and find useful work for the now unemployed bioweapons scientists. What would happen to those efforts if we lose Russian cooperation? One U.S. scientist who was involved in discussions with the former Soviet Union bioweapons scientists thinks that the Russians may be restarting bioweapons research. His concern: he formerly had access to old program facilities and engaged in frank discussions with the scientists; those discussions now are guarded.[25]

Finally, and perhaps most cautionary:

Will America's rivals steer currently legitimate biodefense programs down the new American path, but even more deeply in shadow? In a more recent *Washington Post* article, Leitenberg makes this critical observation: "You can't go around the world yelling about Iranian and North Korean programs—about which we know very little—when we've got all this going on."[26]

It's a disturbing sign, to say the least, that for eight years the Bush administration backed away from the principled, strict interpretation of the Biological Weapons Convention we had adhered to since it became law in 1975.

What Supporters Say

Backers of the Bush administration's massive biodefense spending strategy see biological weapons as such a real and present danger that no amount of countereffort is too large. Gerald Epstein, a former National Security Council science advisor, says that terrorists who even develop a bioweapon can take the next step and scale up production to mass-death levels. Yet as we have repeatedly shown, the gap between a lethal bug and widespread, efficient delivery is enormous, requiring focused scientific efforts of a nation with large resources.

Epstein cites Richard Danzig, former secretary of the Navy and an expert on bioterrorism, as believing that only biological and nuclear weapons can actually pose "an existential threat to the United States." Of the former, Danzig says that in the "extreme but chillingly plausible" event of an assault of sufficiently massive scope, "our national power to manage the consequences of repeated biological attacks could be exhausted while the terrorist ability to reload remains intact."[27]

Epstein similarly avoids even discussing *likelihood* in stating, "Policymakers may be particularly risk-averse when it comes to catastrophic threats. They may not want to play 'bet your country' with even a small risk of a horrific attack."

The key phrase concerning risk here is "extreme but chillingly plausible." Such massive assault capabilities from small numbers of nonstate players—even those, like Aum Shinrikyo, who are fanatically dedicated and extraordinarily well funded—are quite *im* plausible. Neither we nor anyone else can say they are impossible, and the difference between those two, we assert, marks the gulf between *plausible* threats that any highly developed nation should defend against and *all possible* threats that no nation can defend against—unless it has limitless resources. In the current financial crisis, simply indiscriminately spending large amounts of money becomes a security threat itself.

We envision that the real present-day risk is of low-consequence events with some likelihood of occurrence—an attack by terrorists who lace food with ricin, say, or an attack by a terrorist or an unbalanced individual who finds a way to carry out a limited yet terrifying attack with anthrax, as someone already did in October 2001. "Envision" is all any of us can do here, since no one can muster actual probabilities for these events. As we argued earlier, there is really little evidence—and none of that direct—that a bioweapons attack can cause massive casualties.

Finally, compare Epstein's emotionally charged "bet your country" with what we see as a far more realistic appraisal by Admiral William J. Crowe Jr., former chairman of the Joint Chiefs of Staff. "These terrorists cannot destroy us.... We are a country of nearly three hundred million people with infrastructure spread across a nation that has the fourth largest landmass in the world. This is not thermonuclear war we are facing.... The real danger lies not with what the terrorists can do to us but what we can do to ourselves when we are spooked."[28]

However, our arguments are not simply against raising unrealistic fears. In chapter 10 we will attempt to show that money spent on countermeasures for natural infectious diseases will apply directly to biodefense, so regardless of odds that are so impossible to define, we certainly are not playing "bet your country." By contrast, BioShield 2004 restricts funding to the purchase of developed countermeasures that have little commercial potential, so the countermeasures can have little impact on common natural diseases—where commercial potential surely exists. Unlike a broad focus including natural diseases as well as biological weapons threats, BioShield's restrictions force us down a one-way street.

Off in the Wrong Direction

U.S. scientists now are working in force on countermeasures—antibiotics, antivirals, antidotes, vaccines—for bioweapons real and imagined. How do many of them go about their investigations? They have abandoned the safe experimental models they have used for decades, moving their work into deadly microbes. They need to do this in order to up the odds of getting funding, and they must get funding in order to continue working.

"Follow the funding" is the rule of the game in American science, from the mundane to the cutting edge. And the game has become ever more serious. According to David Ozonoff, professor emeritus at Boston University School of Public Health, "The funding situation is so bad now, less than 10% of applications are being funded.... We've never really seen anything quite this bad in the NIH system."[29]

He notes that young scientists are having a rough time getting under way and those coming up for tenure can't get the grants they need to secure that requisite for continuing their academic research. "They are desperate to find money wherever they can," he said. "Some of them are very, very good.... I think going after this biodefense money is one strategy among many for trying to keep your head above water."[30]

Ozonoff's observation is supported by two recent reports from major universities and other institutions [31] decrying the grave consequences for American science from years of flat funding from the National Institutes of Health. Underscoring their message, Joshua Boger, president of the biotechnology industry group BIO, warned, "You can lose a generation of researchers pretty fast—in five or ten years. You create such a discouraging atmosphere that they just go somewhere else instead of academic research. We don't have to lose 50,000 researchers, just 50 really good ones. Once it happens, we won't get those people back."[32]

Let a Thousand Lethal Flowers Bloom

With few exceptions, the news-making, groundbreaking science in America is done in university laboratories by tenured or tenure-track faculty and their students working on U.S. government grants. Most funding in the life sciences comes from the National Institutes of Health, another portion from the National Science Foundation, and the rest scattered among other agencies and companies.

Historically, the bacterium *Escherichia coli* has been the workhorse of molecular biology research. A usually harmless denizen of the planet, *E. coli* lives everywhere from pond scum to the human gut, and most strains are so benign that we are unaware we carry pounds of these bacteria quietly colonizing our large intestines. It is the most-studied, best-described "bug" in the world. Some *E. coli* strains can cause human intestinal illnesses, but they are like bad dogs in the microbial kennel. *E. coli* is so safe your child can use it, literally. Web sites coaching junior high and high school students in science fair projects abound with interesting experiments they can conduct using *E. coli,* even offering readymade mail-order kits.[33]

Obviously, for research in *E. coli* you don't need high-biocontainment labs like those proliferating throughout the United States for bioweapons agent research. So why do scientists need such labs? Because they choose to work on dangerous agents to obtain readily available funding. What agents are prescribed, then, and why? And what is going on in the labs that are popping up in our neighborhoods?

Research is now being conducted in some of the most dangerous pathogens known, even when scientists are asking the kinds of questions for which *E. coli* would suffice.

When working with disease-causing germs, lab workers and the public must be protected from disease. Assigning a level of containment for research on disease-causing germs takes into account whether lab workers or people outside the lab or both need protecting. The NIH also considers many other factors in assigning containment levels for specific infectious disease agents: Is the agent deadly? How does it infect? Is it easily transmitted from person to person? Is it already in the environment? Are there preventions or cures?

The BSL-*3* and BSL-*4* labs discussed earlier mark the upper end of the containment system in use in the United States and elsewhere, in which all microbiology laboratories are classified from levels 1 through 4, each level dictating what kind of work can be done there.[34]

BSL-*1* is the garden variety microbiological lab where workers may study *E. coli* and other agents "not known to consistently cause disease in healthy adults." Workers need only to follow basic lab safety practices.

In BSL-*2* labs, standard for most infectious-disease research, workers must wash their hands often, work surfaces are decontaminated frequently, food is not allowed in the lab, and generally procedures are designed to minimize the creation of splashes or aerosols and prevent contact with agents. Since many of the disease agents studied are already present in the environment, there's no need to protect the public from escape from the laboratory. Here's a key point: all infectious bacterial agents and most viral agents require only BSL-*2.* A few exceptions that require BSL-*3* containment are experiments with aerosols of infectious disease agents, work with large quantities of an agent or with particularly virulent agents, and dissecting diseased animals. Many universities have more than one BSL-*2* lab.

At BSL-*3* things get serious. Agents studied in these labs are highly infective and pose a significant risk to lab workers, or they are easily transmitted and pose a significant risk to people outside the lab, or they are lethal. But BSL-*3* – level diseases can either be prevented with vaccines or cured with antibiotics.

The highest level of biocontainment is BSL-*4.* Like BSL-*3,* this level is reserved for highly infective, highly contagious, or highly lethal agents, but with a further big catch: there is no cure for what these agents deliver.

In 2007 testimony to Congress, the General Accountability Office reported that 1,356 BSL-*3* labs in the United States have registered under the Select Agent Rules.[35] We believe many of the labs are new, and every one aims to conduct experiments dangerous enough to require such highlevel containment of its pathogens.

No government agency is sure what most of these labs are doing, or if they are, they are not telling. We estimate several hundred labs are researching and developing countermeasures for bioweapons agents, with 219 labs registering with the CDC to study anthrax alone.[36] That is massive overkill in our judgment.

Why this proliferation of high-biocontainment laboratories? Since 9/11, funding for bioweapons agents research, both basic and for countermeasure development, has increased severalfold. Scientists researching an interesting molecule in *E. coli,* where it can be done efficiently and safely, now must switch to studying the same molecule in a dangerous bioweapons agent in order to boost their chances for funding. Stop and consider that change, and suddenly the proliferation of BSL-*3* labs takes on appropriate gravity.

In addition to multiplying whatever risk is inherent in running an experiment at BSL-*3* by several hundred-fold, the proliferation can only increase chances for theft and sabotage by an insider with murderous intentions, not to mention chances for accidents or opportunities for foolish actions by immature student lab workers. The United States is encouraging and nurturing a thousand poisonous flowers to bloom.

As the number of researchers in biosafety laboratories increases, access to biological weapons agents and training in their skilled use increases. This markedly increases the risk of politically disaffected or mentally unstable lab workers exploiting these agents for hostile purposes.

The number of people working in biodefense has increased perhaps twentyfold in the past decade. Compared to the established government laboratories at the CDC and Fort Detrick, the risk of accidents and thefts is greater in the university-based biosafety laboratories, where inadequate training, frequent turnover of student workers and other personnel, and inadequate oversight from institutional biosafety committees heightens our risk substantially. We contend that risk of accidents and theft has increased well over the twentyfold that would be predicted based on added personnel.

Veteran anthrax researcher Martin Hugh-Jones of Louisiana State University told the *Baltimore Sun* that before 2001, all anthrax researchers "knew each other by first name." Now, he said, "I see a lot of names I've never heard of.... On a probabilistic basis, there's more of a risk of accidents or attacks."[37] The title of this article: "A Plague of Researchers."

Access to biological weapons agents by streams of newly hired, hurriedly screened lab workers increases the risk of terrorism. Men and women in their early twenties may not yet have formed their political ideologies or displayed symptoms of developing mental illness and so may slip by the facility's security screen. Inadvertently, we could be training the next generation of terrorists in handling and weaponizing deadly pathogens, and we could even be providing the pathogens. In contrast, older workers in established government laboratories have a history that has been evaluated by security.

Consider these events at Harvard in the late 1970s, at the end of the school year.[38] As students filled the Science Center to take final exams, a crew in the basement was cleaning up a laboratory area reserved for talented undergraduates to carry out independent research. Coming upon some jars of unknown content, a crew member asked the student working there what they contained. "Nitroglycerine," he answered nonchalantly. Indeed it was. The building was immediately evacuated and the jars were disposed of—very, very carefully. The enterprising young man had spent much of the semester making nitro, apparently on a whim, and he did not understand what all the fuss was about.

If a student "hacker" tomorrow is as intent on developing a biological weapon on a whim, or for sale to a terrorist group or hostile nation, the outcome might be much darker.

A Manhattan Project for Your Town

Perhaps the words of Senator Bill Frist provide a clue as to the scale the U.S. government envisions for biodefense and defense against infectious diseases:

> I propose an unprecedented effort—a "Manhattan Project for the 21st Century" – not with the goal of creating a destructive new weapon, but to defend against destruction wreaked by infectious disease and biological weapons. I speak of substantial increases in support for fundamental research, medical education, emergency capacity and public health infrastructure; I speak of an unleashing of the private sector and unprecedented collaboration between government and industry and academia; I speak of the creation of secure stores of treatments and vaccines and vast networks of distribution; I speak of action, without

> excuses, without exceptions; with the goal of protecting every American and the capability to help protect the people of the world.[39]

The over $50 billion we have spent so far on biodefense has surpassed the $24.5 billion spent in today's dollars on the Manhattan project.[40]

However, to the extent that Senator Frist is talking about defense against natural disease, we agree with his call for an unprecedented effort. But that effort does not require hundreds of BSL-*3* laboratories and perhaps a dozen of the highest-biocontainment BSL-*4* labs.

Nevertheless, let's ponder his analogy. The Manhattan Project was carried out under the closest secrecy in history up to that time, and the work was done in a small number of university laboratories scattered around the U.S., with many new laboratories in the then-isolated mountaintop village of Los Alamos, New Mexico. Its researchers, small in number, were carefully picked top scientists and technicians, working at a time when electronic espionage was in its infancy. Wartime restrictions all but eliminated international travel.

Imagine doing that work in hundreds of "Los Alamos" labs staffed by people only semitrained because of the urgency of their work, all sporting cellphone cameras and harboring a true American variety of unknown and unknowable political leanings and psychological profiles.

And of course, though they did a great deal of work on The Bomb themselves, the Russians still managed to steal our atomic secrets.

Guards, Guns, and Gates: The Select Agent Rules

Biologist Donald Kennedy, the editor of *Science,* generally considered America's leading scientific journal, offered this down-to-earth take on the crunch when science meets government.

First, there's secrecy:

> Scientific culture is by nature oriented toward disclosure. Because the research venture grows by accumulation of information, it depends on the free availability of previous work through publication. Security requirements, on the other hand, often dictate concealment.

Then the many ways people go about their daily work:

> Scientists tend to be exquisitely careful about data but more casual about office order or precise dates, whereas security processes emphasize compartmentalization and strict adherence to procedure.[41]

Of course, that image of security contradicts the attitude of permissiveness we've been speaking of. If the government is nurturing all those poisonous flowers, at least it seems top researchers should be having a field day in a newly *laissez faire* world. But, as Kennedy implies, it's just the opposite, because working with potential weapons agents and others deemed harmful requires a quest through procedural mazes and regulatory mysteries straight out of an espionage novel.

The government's "select agents" may sound like embedded spies, but in this case they are presumed potential weapons that might fall into enemy hands, and getting to work on them is so onerous that many established top scientists will not participate in biodefense activities. That is the new world, post 9/11. They are biological agents and toxins either designated as potential bioweapons or that otherwise pose a special threat to public health, and they include the Category A bioweapons agents we are already familiar with and a host of other human, animal, and plant pathogens and toxins. The current list, compiled by U.S. Health and Human Services and the Department of Agriculture, comprises eighty-two pathogens and toxins.[42]

Before adding a new agent to the list, the government considers such factors as the effect on human health of exposure to it, the degree of contagiousness and the methods of transferring it to people, its potential for use as a bioweapon, and finally, the availability of drugs and immunizations for prevention and cure of related disease.[43]

Many select agents have existed among us forever and would make poor bioweapons for a variety of reasons, so the inclusion of so many on the list is questionable in itself, and for some agents only the more pathogenic strains warrant designation.

If you are a typical university professor/researcher and want access to any of them, be prepared to follow strict security measures, the culmination of several pieces of enacted legislation [44] and the Select Agent Rules themselves. For example:

- A "responsible official"—you, the principal investigator—must register with HHS or the USDA before getting the agents. And your registration will cover only the specific select agents and toxins and precise activities at the exact locations approved. If you want to change any of that, go back and start again.
- All those who have access to the agent must have an approved security-risk assessment. They must submit a form and fingerprints to the FBI and wait months before starting work.
- Before visiting scientists can work in your lab, they must get their own security-risk assessment or have a current one validated for the new location, a process as lengthy as a new security-risk assessment.
- "Restricted persons" cannot legally possess, transport, or ship any select agent. That means anyone indicted or convicted of crimes punishable by imprisonment for more than one year, fugitives from justice, illegal aliens, dishonorably discharged service members, any "unlawful user of any controlled substance," and anyone who has been "adjudicated

as a mental defective" or committed to any mental institution.[45] Wouldn't all this apply to someone in the shipping department, say?

- Required lab security could include stationing guards, limiting access by outsiders, and providing high-tech identification procedures for anyone entering.

And do not screw up. Criminal and civil penalties for the inappropriate use of select agents and toxins include fines, imprisonment of up to five years, or both. Fines can go up to $250,000 for individuals and $500,000 for an institution.

Security Bites Back

But what is some extra paperwork and precaution in the name of biosecurity? A great deal. Several analyses provide powerful evidence that the measures are severely cutting back on laboratory productivity—that is, on the pace of critical research itself. In an insightful monograph, Julie Fischer of the Stimson Center reported that research showed drastic cutbacks in productivity resulting not only from paperwork demands but from required waits of weeks to months for approvals. She noted that some researchers queried reported following a "two-man rule," in which no researcher ever works alone with a select agent. Obviously, that will slow the pace of work and make it more expensive.[46] Meanwhile, at the University of Wisconsin, a longtime animal pathogen researcher observed, "I can't just say, 'You're hired to work on this,' because it has to go through federal approval. That takes routinely from three months up to six to eight months."[47] In the Stimson study, one researcher even feared that "more time would be spent on paperwork than supervising safe research practices."[48]

Picture yourself considering a new job with a choice between two labs, one subject to the new security measures, the other not. The "select agent" employer cannot even guarantee you a position until the FBI okays the application. Even a mistyped date of birth on your security-risk assessment application could hold up that approval for months. And how about the personal information required? Maybe you smoked marijuana in college and you are afraid the FBI will find out if agents interview your old roommates. Will disclosing it up front haunt you for the rest of your professional life? Will you ever be able to get an NIH grant? Why not just take that other job?

Then the cost of required extra building security must be added in. The Stimson survey reported enormous extra costs cited by an official of Louisiana State University, which was noted for its anthrax expertise even before the October 2001 assaults. Michael Durham, LSU's Director of Occupational and Environmental Safety, said, "We put in $130,000 worth of security equipment in our building and then found, under an inspection made, we were still lacking." And the scope was major, the inspectors recommending, for example, that "bollards be placed in front of a building, concrete obstructions to keep someone from going in with a vehicle and blowing the building up. And this is a veterinary school, a veterinary clinic."[49] The absurdity emerges: cement barriers to protect a veterinary clinic where people and animals come and go all day, a building no more likely a terrorist target than your local hospital.

At another institution, the process for creating an inventory was so detailed it included a freezer map. But the inventory itself, of course, required its own security and needed meticulous upkeep "and would still not detect the surreptitious removal of enough pathogen from a single vial to start a new culture."[50]

From the absurd to the self-defeating: newly required security measures were so attention-getting that one researcher who had been working quietly for years in a lab with limited access and secured doors now drew fire from the entire university community.[51] That is certainly a flashing arrow for someone with hostile intentions wanting to get hold of a select agent.

Consider the government's efforts to hide the names of select-agent researchers because of their fear that the information could end up in terrorist's hands. We carried out a search on the U.S. National Library of Medicine's widely used PubMed database to identify published basic and medical research using names for a few Category A agents. As the "Pub" implies, this is a publicly available online database that identifies and summarizes published research around the world.[52] A PubMed search of "*Bacillus anthracis*" brought up titles, author names, and journal citations for 2,687 publications; "tularemia" raised 2,296 publications; "Ebola," 1,016. However, just as the government suggests is essential for security, we uncovered no university or other institutional designations for the scientists. At least not in the first step we took. However, by simply requesting the abstracts of articles, for most publications checked we would instantly have gotten the names of the researchers *and their institutions,* along with brief descriptions of the work.[53]

In cases where the institution name was hidden for protection, anyone interested would simply look up all work by the same author, and his or her home institution would come up, too.

Of course, not all research on a select agent requires the agent itself. In some cases the investigator needs only a few of its harmless proteins; in others, nonpathogenic strains may be used.

The bottom line remains: someone with a routine bioscience background and hostile intent or pathological personality can easily pinpoint laboratories that work on select agents, so all that secrecy is wasted. But is it even necessary? It would not be with simple steps that would provide needed security.

Here is a low-cost, two-step suggestion:

- Culture collections of select agents are labeled only with cryptic numbers and letters—and, of course, the "biohazard" insignia.
- The key to the labeling system is locked away in a different location with access as restricted as need be.

This leaves the dangerous strains "hiding in plain sight." So if approved researchers want a particular select agent for a new experiment, the "responsible official" would either give them the cryptic label name or would actually retrieve the sample. Other microbes in the lab do not need to be labeled cryptically, but they should be. Doing so would make it very difficult indeed for an evildoer to identify the dangerous special agents. At present, the Select Agent Rules require just the opposite, mandating clear labeling on all select agent containers, making their location obvious to anyone in the lab.

What about the idea of keeping the location of select-agent experiments secret anyway, just to be on the safe side? It would be anything but safe. We have already pointed out how unwarranted secrecy could fuel an arms race in bioweapons, but it is at least as vital for people to know what is going on in their neighborhoods. That means work on select agents should be public, but measures to prevent outsiders from getting their hands on a particular culture should be rigorous, and *they* should be secret.

Ironically, the government's heavy-handed security measures in the opposite direction may have caused considerable damage to our biosecurity by bringing to an end some of the very work that was supposed to be encouraged. The Stimson Center's Fischer notes that the CDC "estimated that 817 entities would register under the new select agent rule, but only 350 had registered by the time of the final rule."[54] Many say this drop reflects pathogen research eliminated because it was not worth the trouble imposed. That accords with the Stimson select-agent survey. Fischer said, "All 28 scientists who responded ... indicated that they had eliminated at least one research project involving a select agent in response to the new biosecurity regulations, and that they personally knew colleagues who had done the same."[55]

The government disagrees, claiming that "of the 200 or so entities that transferred or destroyed their select agents rather than registering under the rule, we believe that the majority did so for reasons that do not threaten future research."[56]

But perhaps the most urgent threat is to research into countermeasures against pandemics. Public health experts almost universally agree that timely international cooperation is needed to detect and reduce the impact of an emerging epidemic or pandemic. Collaborations among scientists of all nations are important in this huge effort. Even though flu viruses are not select agents, the Select Agent Rules may prevent or hamstring important collaborations—for example, by disallowing key foreign scientists from even working in U.S. labs possessing select agents—and that interference could critically delay response to an emerging flu epidemic.

Witch Hunt

Professor Thomas Butler was one of the country's top plague researchers, a physician who had learned about the killer disease firsthand as a naval officer, treating it among Vietnamese villagers. Now he was studying the host *Yersinia pestis* in Tanzania, collecting specimens and shipping them back to his lab. Back at his home lab, he reversed the procedure, shipping some specimens and lab work back to colleagues in Tanzania. And that was all he did. What he didn't do was to get the required permits to do any of that shipping, but then, many scientists didn't bother with these, considering all the paperwork a hindrance to their work that netted nothing positive. Before 9/11, no one took much issue with that.[57]

Butler himself set the wheels in motion for what came next. He discovered—frighteningly to be sure—that he could not account for thirty tubes of *Y. pestis* cultures. He took the appropriate action and notified the biosafety officer at Texas Tech University Health Sciences Center, and ultimately the FBI was called in.

That failed accounting cost Butler two years of his life, spent in prison after a jury convicted him on relatively minor charges of financial misconduct but acquitted him of the big ones that had aroused the FBI's suspicion that he was guilty of bioterrorism-related crimes. The financial misconduct charges would have normally resulted in only a slap on the wrist, certainly not jail time.

Agents pursued Butler as though he were the bioterrorist incarnate and many scientist commentators have bemoaned the ensuing trial and its furor for ending the career of an extremely talented, meticulous, and caring researcher. Others say that Butler, however inadvertently, was rightly called to account for errors that in these times could have serious consequences and that the case sent a "clear signal" to all scientists. The clear signal to some scientists is, "Do not begin work on select agents. The risk and hassle are too much."

At least Butler actually made mistakes, by his own admission. Not so the brothers Irshad Ali Shaikh and Masood Ali Shaikh, Pakistani immigrants to the United States and physicians with outstanding records of achievement and of humanitarian effort. One brother was health commissioner for Chester, Pennsylvania. The other was that city's epidemiologist. Both have made major contributions to American efforts in Iraq and Afghanistan, working on government assignment. Nor were any mistakes made by Asif Kazi, who had been the brothers' childhood friend in Pakistan and was Chester's city accountant.

In the days after the anthrax attacks, the FBI visited all three, reportedly acting on a bad tip.[58] But "visited" doesn't quite do it. An FBI SWAT team burst into the home the Shaikh brothers shared, awakened a wife who had been sleeping, and handcuffed her at gunpoint. The SWAT team then smashed a downstairs office rented by the city's AIDS Care Group and handcuffed a carpenter working there. They conducted a similar raid on Kazi's home a few blocks away, bursting in as his wife prepared breakfast and holding her at gunpoint as they carried out computers and files.

Finding nothing did not dissuade the agency. For years after that November day, Irshad Ali Shaikh was blocked by the FBI from obtaining a federal contract with the CDC. Immigration officials canceled his scheduled interview for U.S. citizenship. And whenever he returned to the United States from abroad, federal agents escorted him off the airplane and interrogated him for hours.

Their ordeal "ended" with the FBI's conclusion that the anthrax letters had been the work of scientist Bruce Ivins, whom we will discuss later. The three—who have never criticized the FBI for the investigation—were lucky in that their home city formally supported them as innocent, and they still hold their city jobs. Now a congressman who was mayor of Chester is trying to get them officially exonerated. Nevertheless, Kazi noted in August, 2008, "As it stands, the investigation is still open.... If I go and look for another job, they will check my record, and who knows what would happen?"[59]

Finally, the bizarre story of Steven Kurtz, protest-artist and professor of art at the State University of New York, Buffalo. His wife died from heart failure in her sleep one night and his home was raided the next day by FBI agents decked out in full biohazard gear. His art materials and even his wife's body were impounded. Make sense?

The FBI's paranoia had been set off by a police report. According to a *Washington Post* article written a few weeks after the raid, after Kurtz called 911 to report that his wife had died, police and emergency workers found "vials and bacterial cultures and strange contraptions and laboratory equipment."[60]

That resulted in two days of questioning by the FBI in a hotel room. However, Kurtz's bacterial cultures were an element of his politically oriented art. A vocal social critic, he and his wife of twenty years were among founders of the Critical Art Ensemble, a nationally known group of politically engaged artists, and he was a practitioner of what has become known as "bio-art."

But the seizures and grilling represent the benign part of this story. Kurtz had been developing a new project, and the harmless bacterial cultures he was growing were to simulate anthrax and plague attacks. He'd gotten his samples from a colleague, geneticist Robert Ferrell at the University of Pittsburgh Medical Center, who had ordered them for him from the American Type Culture Collection. That sparked fears that the two were part of a bioterror ring, but once state health department officials learned the bacteria were harmless and that Kurtz's wife had died of natural causes, they dropped all charges.

Not so the Justice Department. In 2004, it charged both men with wire fraud under the Patriot Act. Now they faced twenty years in prison. Ferrell had only been doing a favor for a friend, and until a few years ago Kurtz wouldn't even have needed the favor. He could have ordered the bacteria directly from the American Type Culture Collection, no questions asked—but not in this climate of fear. Artists angrily protested that Kurtz was singled out as a critic of government science policy—in fact, his art often attacks genetic modification of foods and is critical of modern biotechnology.

As the case dragged on, the leading journal *Nature* agreed with the artists and urged scientists to support Kurtz. "As with the prosecution of some scientists in recent years, it seems that government lawyers are singling Kurtz out as a warning to the broader artistic community," *Nature* editorialized, months after the case had begun. "Kurtz's work is at times critical of science, but researchers should nevertheless be willing to support him.... Art and science are forms of human enquiry that can be illuminating and controversial, and the freedoms of both must be preserved as part of a healthy democracy—as must a sense of proportion."[61]

A sense of proportion, indeed.

Four years after the hunt began, Ferrell, his health failing, pleaded guilty to lesser charges and paid a $500 fine.[62] In April 2008, Kurtz was found not guilty of federal bioterrorism charges, and his ordeal ended.[63]

In other cases, we'll see that federal prosecutors seem blasé about select-agent rule violations that actually have caused injury and put the public at risk at a number of institutions.

Our Conclusions

America's Select Agent Rules and related security measures so impede scientists' ability to carry out research that many experienced researchers are abandoning biodefense research. FBI persecution of scientists who have done nothing wrong and of others whose errors have been blown out of proportion by zeal, paranoia, or a combination of the two compound the problem. We long for biosecurity as well, but the net effect of our frenzied, draconian attempts to make us more secure from hostile use of select agents are taking us down the opposite road.

The intangible costs of our actions may be the most damaging to our biosecurity, most notably in the effect on international collaborations and cooperation that could hurt international public health efforts and hinder the rapid response needed to thwart emerging pandemics.

Boston University's David Ozonoff adds another perspective: that we are attacking ourselves. "Bioterrorism to me is analogous to an autoimmune disease. We did it to the Soviet Union, we bankrupted them in the arms race. Now,[al-Qaeda is] going to bankrupt us on the biodefense stuff."[64]

CHAPTER SIX

Dangerous Acquaintances

> I have lost all sense of smell, and have the broadest range of allergies of anyone I know. I can't eat butter, cheese, eggs, mayonnaise, sausages, chocolate, or candy. I swallow two or three pills of anti-allergy medicine a day—more on bad days, when my sinuses start to drain. Every morning, I rub ointment over my face, neck, and hands to give my skin the natural lubricants it has lost. The countless vaccinations I received against anthrax, plague, and tularemia weakened my resistance to disease and probably shortened my life. A bioweapons lab leaves its mark on a person forever.
>
> KEN ALIBEK, BIOHAZARD

Reading through the eight-year history since the 2001 anthrax-letter attacks, the lethal mailings seem to have come out of the blue, without anticipation. But as this 1999 *New York Magazine* cartoon by artist Danny Hellman reminds us, the use of mailed anthrax spores as a homemade bioweapon had been anticipated by emergency planners for many years, and for that long hoaxers had been mailing envelopes of white powder alleged to be anthrax. The hoaxes continue, almost daily. Only once so far was the threat real, and in many ways the scenario mimicked the stereotype: envelopes addressed in crudely written, all-capital letters; messages all the same. The one delivered to Senator Patrick Leahy (D-VT) read:

FIGURE 6.1. Terror by mail: the anthrax letters

WE HAVE ANTHRAX
YOU DIE NOW
ARE YOU AFRAID?
DEATH TO AMERICA
DEATH TO ISRAEL
ALLAH IS GREAT

Before the weeks-long crisis ended, five people were dead and many more were sickened. Who was behind it? If this was a boast by Islamic radical terrorists as a follow-up to the 9/11 attacks, it lacked the one thing nearly all others have had in common: a signature. No individual or group ever has laid claim to an act that truly succeeded in deepening and widening the fear Americans felt in the last months of 2001. Why would a terrorist not claim his act of terror? That said, some still believe that terrorists such as al-Qaeda were behind the attacks.[1]

However, one fact has emerged over the years: the anthrax came from the government's own laboratories, most likely those at Fort Detrick. Almost certainly the perpetrator was an unbalanced or disgruntled lab scientist. FBI investigators say that individual was Bruce Ivins, a senior scientist at Fort Detrick who had cultured the anthrax they are certain was the source of the spores mailed in 2001.[2] Ivins died of an overdose of medication in July 2008, and his wife said she believed he had tried to commit suicide before by similar means.[3] Stories since his death paint a conflicting picture of a man privately tortured by mental illness and alcoholism, yet who was a great friend and solid colleague to those who knew and worked with him.

Many of those friends and colleagues believe that, whatever his troubles, Ivins was innocent of the terrible crime of which he was accused, and they say that other laboratory workers in and far from Fort Detrick may have had access to the anthrax. They point out that all the evidence against Ivins was entirely circumstantial and that the pressures that drove him to his death were inflicted by the investigation itself.

The FBI revealed that Ivins's flask, labeled RMR-1029, contained anthrax Ivins cultivated himself, which contained genetic signatures found only there. The signature matched a sample Ivins sent to an Arizona researcher, who provided it to investigators early in the search. They say the letter-attack spores contained the same, unique signatures.[4]

Does that mean that others—especially those with terrorist intentions—could not have stolen some of the contents from Ivins? Of course they could have. That is why the mystery is not likely to be solved. What it clearly does show is that, *even ordinarily,* hundreds of lab workers do have access to potential bioweapons agents, such as the Ames strain of anthrax. We clearly are not safe from our own laboratories.

For much of the time after the attacks, the FBI focused not on Ivins but on another local scientist, Dr. Steven J. Hatfill. Six years after coming under around-the-clock scrutiny by the FBI and constantly being in the media spotlight, Hatfill was exonerated. In a telling comment on the entire Amerithrax investigation, *New York Times* reporter Scott Shane noted of the Ivins case:

> Some F.B.I. agents were haunted by the Hatfill precedent. Dr. Hatfill, too ... had begun drinking

> heavily as he came under scrutiny. He, too, had grown depressed and erratic under the F.B.I.'s relentless gaze.
>
> What if Dr. Hatfill had committed suicide in 2002, as friends feared he might? Would the investigators have released their evidence and announced that the perpetrator was dead?[5]

As we'll see, Hatfill was far from the only scientist hounded by suspicion during the intensive, years-long investigation, and whether that was the unfortunate byproduct of resolving a critical public safety issue or the direct result of a witch hunt may be argued for as long as the true culprit's identity.

One lesson emerges with crystal clarity: the proliferation of high-level BSL-*3* and BSL-*4* laboratories radically increases our risk of a deadly bioweapons attack—most likely through the actions of disaffected or deranged scientists or other laboratory workers, but possibly through theft by terrorists. That proliferation in turn is aimed at countering a very lowprobability threat that terrorists could engineer, produce, and deliver such weapons on their own.

Why would they bother? The easiest way for terrorists to cause massive panic and death with bioweapons would be to place one graduate student or postdoctoral fellow in one of the hundreds of university labs in the BSL-*3* or BSL-*4* category in order to replicate on a larger scale what someone already did.

The World of the Possible

As the number of high-biocontainment laboratories handling dangerous pathogens mushrooms, so does the opportunity for deliberate wrongdoing. And so do the odds of a whole host of accidents just waiting to happen. Human error encompasses all sorts of slipups that often seem utterly implausible—until they occur. Laboratory security can never be designed brilliantly enough to be unbreakable.

Maybe this seems needlessly alarmist, as though we were unrealistically projecting a future chock-full of dangerous experiments unsafely conducted, even as we criticize the government for fearmongering. In fact, we are simply projecting the recent past onto the near future, and it is a past seemingly guided by Murphy's Law:

- Three plague-infested rats were unaccounted for in a BSL-*3* laboratory in the heart of a Newark, New Jersey, residential area.
- A leaky "foolproof" aerosol chamber infected three lab workers with tuberculosis.
- Tularemia sickened three researchers who thought they were working with a benign strain of the rabbit fever agent.

Finally, an episode involving Ivins that we find ominously instructive and will detail later:

- A colleague's dust-covered desk caught a researcher's eye and turned out to be laden with anthrax spores.

The Council for Responsible Genetics counted more than seventy accidents relating to work with dangerous pathogens from 2002 into 2007.[6] No one knows how many lab accidents go unreported, but the true number may be far greater. Since the council's tally, a number of recent accidents have been reported by the Associated Press and the Sunshine Project.[7]

But beyond simple accidents lie hazardous events far harder to predict, with far greater potential to cause harm. Careful experiments have unintended—sometimes truly unforeseeable—results. Perhaps most important, well-motivated efforts to understand and even create new microbes, no matter how carefully carried out, dramatically up the odds of terrible consequences. Combining those two just a few years ago led to a profoundly cautionary tale.

The Best Laid Plans of Men for Mice

Australian scientists were experimenting with a fairly benign mousepox virus strain—benign because it does not affect humans or other mammals at all and causes the equivalent of a common cold even in mice.[8] Some days after the mice were infected with mousepox, on what should have been a routine day in the lab, the experimenters were stunned to find the mice dead. This "common cold" had killed every one. But of course, what the researchers had injected was anything but common, and therein lay an important discovery.

The long-term goal in these genetic manipulations had been to curtail a mouse population devastating Australian crops for years. The researchers believed they had found a way to do just that, by rendering whole populations of female mice sterile. The mousepox virus they were using had been engineered to carry a gene for a mouse "eggshell" protein. The idea was for the injected mouse to react with an immune attack against both the virus and the eggshell protein in order to obstruct eggshell production. If the animal's own immune system halted eggshell production, the mouse would be sterile.

To increase the strength of the immune attack, the researchers inserted another gene that boosted immune response, called interleukin-4 (IL-4), into the mousepox virus. Sterile but healthy female mice now would be able to mate with any number of potent males but they would be unable to produce any offspring. Better still, they would pass on both the mousepox virus and the two genes linked to it through the wild mice of the countryside.[9]

Checking up on the mice on day six after infection, the scientists noticed their feet had begun to swell where the virus had been injected. On day eight, all of them were dead. Fertility was moot.

How had the animals been killed in this strange outcome to an experiment that should merely have caused mild infection? The investigators decided to inject the preparation into mice that had been immunized against mousepox. In other words, these mice should have been protected from any infection. Strange became ominous. Sixty percent of the protected mice died as well. It did not take long for the scientists to realize the meaning of the unexpected outcome—and the potentially lethal implications for humans.

To render the mice sterile, the experimenters had needed to deliver the genes for a fertility brake deep in the mouse's DNA. To get such proteins working, they attached their vanishingly tiny genes as the payload on a "rocket," that delivery missile being mousepox virus. Like all viruses, this one hijacked the mice's genetic machinery so it made the IL-4 as well as the eggshell protein.

Over time, they expected to find the mice simply under the weather for a few days but permanently sterile, passing along to other mice both mousepox and IL-4–borne sterility.

Here is what turned two harmless ingredients into a deadly pathogen and triggered the researchers' alarm. Unexpectedly, IL-4 actually disabled the rodents' immune systems, and for mice without immune response, mousepox was lethal. Even more ominously, this missile nearly wiped out the so-called immune memory that vaccines produce to do their work, rendering vaccination worthless for 60 percent of the inoculated mice. Of course, it was not the future of mice that had the scientists on edge, but the implications for humans.

Mousepox is a cousin of cowpox. Cowpox is akin to smallpox—akin, but distant enough that it was used two hundred years ago to create the first vaccine against smallpox. Cowpox causes only mild symptoms in humans but creates immunity against one of the world's deadliest viruses. Vaccinia, a relative of cowpox virus, is used today to vaccinate against smallpox. Could the procedure that turned mousepox deadly for mice be adapted to turn cowpox deadly for humans?

Or what if a molecular biologist using the string of cookbook techniques these researchers had employed came up with a protocol for inserting human IL-4 into smallpox? Smallpox is considered among the worst diseases ever, with a 30 percent mortality rate. This new combination could kill every last victim, and immunization would be all but futile. Carrying out the genetic manipulation itself would not be difficult, though only if the smallpox virus were available. So far, as far as we know, it is not in a rogue nation's armory or a terrorist's hands, the only stocks held securely by the United States and Russia.

The Australian scientists notified their government of the potential for biowarfare development they had discovered. Australia then alerted the biosecurity community to this deadly new biowarfare "missile"—a weapon so far not deployed in the field but certainly out of the box.

The mousepox case raises critical questions scientists and policy makers must begin to answer now.[10] Should scientists be required to alert others to dangerous accidental findings? No such notification is mandated anywhere now, to our knowledge. Should such findings be published or kept secret? Here the road forks, with powerful arguments for each choice made by leading scientists.

Mark Buller, a virologist at St. Louis University, is a firm believer in the strategy of genetically engineering deadly organisms in order to understand their mechanisms and to learn to defeat them. He took up the gauntlet, repeated the Australian experiment, and extended the work to find countermeasures to the mousepox IL-4 pathogen. Buller and his team designed a two-drug combination they hoped would defeat it, but to date there is no published report on whether those experiments have been successful.

Meanwhile, Buller's team went a step further, engineering a mousepox strain that killed 100 percent of the vaccinated mice even when they were also treated with an antiviral drug.[11] And again they began seeking a countermeasure.

"All this is out there," Buller said of such research. "There are cookbooks easily attainable on how to make this stuff."[12]

However, Ken Alibek, the former Soviet biological weapons leader, believes such work should be kept secret so that terrorists and rogue nations will not learn the secrets of genetically engineering weapons.

Slippery Slope

Mousepox is related to cowpox, cowpox is related to smallpox. ... Even if smallpox cannot be stolen or illegally bought, cowpox is easy to come by and it infects humans. It's easy to carry the mousepox analogy forward to imagine making cowpox itself deadly.

Attempting to do that was Buller's next move, inserting that infamous mouse IL-4 gene into cowpox. The experiments were carried out at the BSL-*3* level of containment even though, Buller asserts, the IL-4 protein is species specific, and he believes there is no chance that even if the cowpox were to infect a human the mouse protein would do its deadly work.

But a member of the original Australian research team strongly disagreed with doing this work. Ian Ramshaw of the Australian National University in Canberra said, "I have great concern about doing this in a pox virus that can cross species."[13] Ramshaw agreed that viruses containing mouse IL-4 should not be lethal to humans, but, he noted, recombinant viruses can do strange things. The Australian team continued to hunt for countermeasures for its own IL-4 mousepox, eventually finding an antiviral drug that delayed onset of death for the mice, though it did not prevent it.[14]

One bit of good news. Lethal as the Australians' mousepox IL-4 combination was to the experimental mice, it was not contagious. Uninjected mice did not catch the deadly disease. However, before we are too cheered by that ray of hope, let's remember that if some form of this combination could be delivered against humans in an aerosol, the fact that it is deadly but not contagious could be an advantage for an attacker. It would mean the weapon would not "blow back" on those delivering it, and they could then occupy the infected territory. Anthrax, however, is probably still far superior as a bioweapon, because its spores are so stable. Creating new bioweapons that are deadlier than what nature has already contrived is not easy. Fortunately.

Buller may have moved this research from St. Louis University to the secret labs at Fort Detrick, since new information has not been forthcoming and it is hard to believe the work was simply dropped. Does this work violate the Biological Weapons Convention? It certainly constitutes development of a new biological weapon, and that could be interpreted as a violation even if meant for defense. But that is not clear from the language of the convention itself. Consider this critical statement:

Each party to the convention "undertakes never in any circumstances [to] develop, produce, stockpile or otherwise acquire or retain ... microbial or other biological agents, or toxins whatever their origin or method of production, of types and quantities that have no justification for prophylactic, protective or other peaceful purposes."[15]

Under the usual interpretation of this language, small amounts of *existing* agents can be retained and researched for defensive purposes. Countermeasure development would constitute such a protective purpose, but that does not speak to the creation of new agents for investigation. And whether or not it would be in violation of the BWC, the Australian experiments and those that ensued could easily be repeated in any nation that has routine genetic engineering skills—Iran, Cuba, and others—and that could be the harbinger of a horrible biological weapon.

What's the solution? The bottom line is to provide complete transparency in any such efforts. First, research on IL-4–pox combination viruses should continue, to confirm beyond doubt that they are not contagious and to develop countermeasures if needed. But we urge caution appropriate to the virus under study: research into mousepox has been going on successfully; research into cowpox should be conducted very carefully, and research into smallpox should not be carried out at all.

Here is the most important caveat: all this research should be conducted "in broad daylight," completely transparently, so that other nations have no doubt of anyone's intentions. As countermeasures are developed, they should be made available universally, both for humanitarian reasons and as a disincentive for anyone to develop an IL-4–pox virus weapon.

How does this position compare with the original aims of the Bio-Threat Characterization Center to develop new microbial agents in order to defend against them? The IL-4–pox viruses now are a well-known potential threat, not an imagined development. We see no justification for creating imagined new agents in order to defend against them. However, now the new director of the National Biodefense Analysis and Countermeasures Center, J. Patrick Fitch, claims that the BTCC will not develop new agents just to see if it can be done, and that should resolve that concern. Equally important, he has appointed an outside panel of seven scientists to make sure the BWC is not violated.[16] Since developing new agents is forbidden by the BWC, this should effectively halt some BTCC activities that may be in violation of the BWC.

Death and Resurrection, Frankenstein Style

It was an apparently mutated avian flu virus that swept the globe in 1918 and killed an estimated forty million people before it disappeared, presumably retreating into the darkness from which it had emerged. At that time, scientists had absolutely no way of understanding the mechanisms that had turned an anonymous virus into such a killer.

Unlike later serious flu outbreaks, the 1918 strain killed mainly young people, filling their lungs with fluid and destroying lung tissue. Why the young were killed more often by the virus is not known, but it has been speculated that older people may had been exposed to a milder related virus many years before and thus had immunity to the 1918 strain—as cowpox-infected milkmaids did to smallpox. Although its return would be a frightening prospect, this time we might develop a vaccine and understand in advance its pathogenic mechanisms in order to develop treatment. But how would we learn the modus operandi of a century-past killer? For better or worse, that answer has already been given: In an endeavor spanning half a century and thousands of miles, the 1918 flu virus has been brought back to life in the laboratory.

In 1949, a young Swedish scientist studying microbiology at the University of Iowa made it his life's mission to accomplish that goal.[17] Twentysix-year-old Dr. Johan Hultin and his wife drove up the Alaska Highway in search of the burial ground of some victims who had been interred in the permafrost, where he believed their bodies would be well preserved. At the tiny settlement of Brevig Mission, the viral strain that Hultin called "the most lethal organism in the history of man" had truly lived up to the claim. Two days after the outbreak, seventy-two out of eighty residents were dead. It would take Hultin two years before he could return and, with the permission of native elders, open the mass grave marked by two crosses where missionaries had buried the bodies of those seventytwo victims. He removed lung tissue from four bodies and returned to Iowa, but he failed to revive the virus. In retrospect, of course, it is lucky he failed, since sophisticated biocontainment labs did not yet exist. He might have unleashed a second coming of the deadly virus before there were the means to contain it, let alone to develop countermeasures.

Then, in 1997, living in San Francisco, Hultin read an article in *Science* written by Dr. Jeffery Taubenberger of the Armed Forces Institute of Pathology in Washington, D.C. Using the new techniques of genetic engineering, Taubenberger and colleagues had developed a method to study the 1918 virus using tissue samples of two young soldiers killed by it. But they did not have enough viral material to work on.

Hultin leapt to the challenge. Using his own savings, Hultin, now seventy-two years old, flew back to Brevig Mission and, with renewed permission, again opened the mass grave. He returned with the lungs of an elderly woman, almost perfectly preserved.

"The only sample we now have is there because the elders let me go back into the grave again," he told a writer for the Alaska Science Forum. "They gave us the opportunity to do something good—not just for themselves but for the whole world."[18]

The rest, indeed, is history—and it is a highly controversial one. Bringing the 1918 flu virus back to life is precisely what science has accomplished, but unlike Hultin and some leading scientists, we and many of the nation's top infectious disease specialists believe it was a terrible mistake. Now any lab of talented virologists anywhere can apply to the CDC for a sample to study themselves.

The rift between the two sides of this issue was further sharpened when *Science* decided to publish the entire viral sequence. Philip Sharp, a noted MIT virologist, applauded the achievement in an editorial accompanying the publication. "The good news is that we now have the sequence of this virus, perhaps permitting the development of new therapies and vaccines to protect against another such pandemic," he said. "The concern is that a terrorist group or a careless investigator could convert this new knowledge into another pandemic," setting out the two sides of the controversy and adding, "The dual-use nature of biological information has been debated widely since September 11, 2001."[19]

That phrase "dual use" will play a major role in our discussion later, but Sharp came down on the side of increasing scientific knowledge. "I firmly believe that allowing the publication of this information was the correct decision in terms of both national security and public health," he wrote.

Publishing the sequence will allow safe research on the virus's genes, but research on whole infectious virus would be a different story.

Dr. Donald Henderson, who was a leader in the World Health Organization smallpox eradication campaign, said, "The potential implications of an infected lab worker—and [of] spread beyond the lab—are terrifying."[20]

The CDC, which controls the U.S. viral cultures, says it will distribute them to legitimate researchers with BSL-*3* "plus" laboratories. That BSL-*4* labs should have been required but were not shocked Rutgers microbiologist Richard Ebright, who said that "only the most complacent, or most ethically challenged, researcher could fail to understand this."

Inflaming critics even more, the CDC announced that it was classifying the 1918 pandemic flu virus as a select agent. "A man-made select agent and a new threat from which we must defend ourselves," said Edward Hammond, director of the U.S. office of the Sunshine Project. "The 1918 bugs are out of the jar and will never be put back in, unfortunately. Way to go, biodefense program!"[21]

One leading 1918 virus investigator responded that critics did not understand the full extent of his team's precautions. Yoshihiro Kawaoka said his workers get regular flu vaccinations, which protect mice against 1918like flu viruses, and they also take the antiflu drug Tamiflu as a preventative. Work by his group and by a federal labs showed that this antiviral "works extremely well" at protecting mice against 1918 like flu strains, he said.

Such confidence in the efficacy for humans of a drug that protects mice is downright scary, all other issues aside. Most drugs that *enter* clinical trials have been tested in mice, where they worked. But nearly half those drugs—40 percent—turn out to be ineffective in humans. Tamiflu has been approved by the FDA for some antiviral treatment, so it may work against the 1918 virus in humans, but we do not know that. Worse, suppose a lab worker were to become infected and transmit it to his or her spouse, who would not be taking Tamiflu. The spouse could come down with a full-blown infection of the lethal flu, starting a pandemic even as vaccinated lab workers remain healthy.

As for the killing power of this reconstructed virus, scientists who tested it on monkeys were startled that the animals' lungs were destroyed in just days as their immune systems went into overdrive, according to their journal article in *Nature.*[22]

According to a BBC interview with the *Nature* authors, "Symptoms appeared within 24 hours of exposure to the virus, and the subsequent destruction of lung tissue was so widespread that, had the monkeys not been killed a few days later, they would literally have drowned in their own blood."[23]

Viruses are so basic that they exist on the thin margins of life, hijacking an invaded cell's genetic machinery to reproduce more virus. How can such a primitive construct destroy lung tissue? Indirectly, by eliciting a massive immune response from the infected host. It is that hypercharged, inflammatory immune response that kills. Victims' lungs are destroyed by their own immune systems.

The BBC program quoted several experts who ultimately lauded the reconstruction, including the *Nature* lead author, largely for its potential to shed light on similar viruses. Given the worldwide fear of a new pandemic emerging out of such avian flu viruses as H5N1, the argument sounds convincing. But Ebola researcher Jens Kuhn argued at the time of its publication that the *Nature* article may have received so much attention mainly because of the dangerous pathogen involved with the research rather than the actual scientific accomplishment. He told us that the researchers seem to argue that their experiments are the only way to shed light on how the 1918 influenza virus did its deadly work and that it is the gateway to understand avian influenza. But, Kuhn said, "I still don't believe the resurrection was absolutely necessary.[24]

A critical point is that we can shed a great deal of light on how a virus works by studying its genes in isolation or by studying a "defanged" version of the virus itself.

Or why not study a virus that is similar but nonlethal? This would allow work to move forward far more quickly, because it could be carried out at a lower biocontainment level. Some of the work on the intact infective virus is actually being done at BSL-*4*, though we believe that all of it should be. Because such containment requires so many elaborate precautions, research moves very slowly. And working directly with the deadly 1918 virus introduces new risks for what may turn out to be little gain.

What, then, about the argument that we need a treatment and a vaccine in case the 1918 flu itself reemerges? A far more likely scenario for such an outbreak is that an infected researcher would spread it, as we suggested.

To Keep or to Kill?

Eradication of smallpox from the entire planet was a monumental achievement of human dedication and international cooperation. Bubonic plague was far more devastating centuries ago in its concentrated periods of time. The 1918 influenza epidemic arrived, killed, and disappeared quickly. But through two-thirds of the twentieth century, the images of smallpox victims with swollen bodies covered head to toe in pustules haunted those who treated or saw them. Finally, in the 1970s, international teams of public health physicians, nurses, and other health workers were able to target vaccinations to enough people in the poor countries where smallpox remained that the WHO could declare smallpox exterminated. Using a clever "ring" vaccination strategy, they inoculated all the people around known victims, which was necessary since there wasn't enough vaccine for everyone. They first identified those infected by showing pictures of victims to people as they canvassed village by village, then immediately vaccinated all those near anyone infected.

At Russia's Vector facility, a former Stalinist prison, scientists study deadly Ebola, hantavirus, and Marburg viruses in a maximum-containment setting. But only a very small group is allowed near Building Six, the smallpox "cellblock." Across the world in Atlanta, Georgia, the CDC requires its researchers to enter the smallpox containment facility two at a time, each using a different key, under close security surveillance.

For many scientists, such security is not enough. Called "destructionists," they want the remaining stocks of smallpox eliminated. Opposing them are equally ardent researchers—"retentionists"—who are trying to learn what makes the virus tick, what makes it kill, and how a new oral vaccine might be developed. Retentionists further argue that terrorists or rogue nations may have bought or stolen smallpox during the collapse of the Soviet Union, so the world must be prepared for its reemergence as a weapon.

Remember, however, that this argument was made during the run-up to the invasion of Iraq by the Bush administration, and the consequences of subsequent vaccination were in some cases lethal. Anticipating that Saddam Hussein might launch an attack using smallpox as a weapon, the president inaugurated a universal vaccination campaign, beginning with first responders. The result? Up to eight of those vaccinated may have died of complications from the vaccine.[25] The program stalled, "because few hospitals and individuals agreed to participate" in what was a voluntary vaccination campaign, "partly because of perceived risk but mainly because of skepticism about the need."[26] Of course, the aftermath of the Iraqi invasion revealed that Saddam Hussein no longer had biological or chemical weapons of any sort, and there never had been evidence he harbored smallpox.

Victor Sidel, former president of the American Public Health Association, and his colleagues have strong words on the Bush universal vaccination campaign. "In the absence of any smallpox cases worldwide or any scientific basis for expecting an outbreak, these deaths and other serious adverse events are inexcusable."[27]

The new smallpox research also has some strange twists. Dr. Peter Jahrling, a major proponent of new research to develop countermeasures to smallpox, managed to infect macaques with the disease. He injected the monkeys with high concentrations of the virus, developing an "animal model" for smallpox.[28] On the one hand, that is procedurally required, since humans cannot be given smallpox as a means of developing countermeasures. But that move brought sharp criticism from some scientists, who argued that smallpox infects only humans and has no animal reservoir, so infecting a new species could only be dangerous. Nevertheless, Jahrling's work was praised by many colleagues, because he succeeded in producing an animal infection that could be closely studied with the latest techniques of molecular biology.

Pushing smallpox research in new directions has raised major worries even among scientists who are not committed to destroying the virus. Smallpox fighter Henderson said of plans to genetically engineer the virus for study, "I'd be happier if we were not, and the simple reason is I just don't think it serves a purpose I can support. The less we do with the smallpox virus and the less we do in the way of manipulation at this point I think the better off we are."[29]

Dr. Jack Woodall, a leading virologist who was a founder of ProMED, the premier online reporting system for outbreaks of emerging infectious diseases, spoke even more strongly against this research: "However great the theoretical interest in studying smallpox virus with the latest molecular techniques, the risk of an escape is unacceptably high."[30]

And escapes have consequences. He added: "Recent laboratory accidents have shown that however secure the laboratory facilities, laboratory workers have become infected with SARS [severe acute respiratory syndrome] virus and tularemia bacteria, and in fact the last recorded outbreak of smallpox began with a laboratory infection in England.... Surely humankind was given self-awareness in order to help ensure its own survival, not to play Russian roulette with it."

The debate understandably reaches far beyond researchers. Developing countries are scared to death. They have no way of acquiring enough vaccine or marshaling resources to vaccinate their own citizens, so a smallpox epidemic would sweep through their populations like wildfire. These are dangerous experiments when done under the best of circumstances—in BSL-*4* laboratories in public view—but carrying them out in the dark adds a far greater element of risk. That, of course, is where the United States is headed now, as more and more of this research is conducted.

Paranoia leads to secrecy, and secrecy fuels paranoia among nations. But secrecy can create dangers all its own. When dangerous activities are carried out undercover, two disincentives to report accidents arise: the news stories resulting from disclosure would lay bare the dangerous work leading to an accident, and a lab's neighboring citizens would be outraged over the hidden dangers they had unknowingly faced.

Never was the deadliness of secrecy better demonstrated than in 1979 in the Russian city of Sverdlovsk (now Yekaterinburg). A plume of aerosolized anthrax escaped from a Soviet military facility in one of history's worst biological accidents. Eventually sixty-eight people died, but many of them could have been saved—just as most of the Amerithrax victims later would be—had the government not covered up the incident and lied about the cause of illnesses that suddenly had befallen hundreds of people.

Researchers Matthew Meselson and Jeanne Guillemin were part of the investigating team that brought the Sverdlovsk disaster to the world's attention. Guillemin recalls, "Not aware of anthrax development and production at the military facility, local public-health physicians took more than a week to identify the disease, by which time many of the 68 victims were already dead or dying. Moscow public-health officials quickly blamed the outbreak on consumption of infected meat. Those who died later in the outbreak might have been saved by treatment if the risks of exposure to an anthrax aerosol been known to the community."[31]

The 2003 SARS epidemic was fueled by the Chinese government's initial denial of it. Fortunately, the government admitted to the source of the viral outbreak in time to begin controlling its spread. However, if such an outbreak were caused by a select agent today, the outcome might be very different. As Guillemin points out, "Government denial would persist, supported by organizational secrecy that corrupts communication and costs lives. This predictable state reaction offers the best argument [for] transparency in biodefense projects."[32]

Had it not been for a massive international effort and complete disclosure by the Chinese, the world could have faced a SARS epidemic on the scale of the 1918 flu. We barely dodged a bullet. Now many researchers believe that the next SARS outbreak will come from a research laboratory. Indeed, three laboratory workers have been infected by SARS in Taiwan, Singapore, and Beijing since the outbreak was contained. Fortunately these infected workers did not transmit it to others. Asian laboratories are often lax in their biocontainment practices—an uncomfortable truth, but the truth all the same.

Secrecy can have similar effects right in our neighborhoods, where we live and shop. In 2004, chicken flu broke out at farms in Delaware, Maryland, and Virginia, the region called Delmarva. As reported on the infectious disease specialists' ProMED forum, the "poultry producers asked lawmakers ... for legislation to conceal the identity of infected farms, saying they want to avoid panic embargoes by overseas purchasers."

Such cover-ups are virtually always requested or carried out for "good motives," in this case, clearly economic ones. The bird flu in Delmarva was a weak strain. What if a more dangerous strain emerged, and in a country entirely dependent on tourism for its economy?

An anonymous ProMED moderator who has worked with several governments said on the list that secrecy always makes a difficult situation worse. Eventually, the information comes out, and when it does "it makes people angry and confrontational." By contrast, the moderator said, "the rapid release of 'bad' news means that the person releasing it has control of that news and journalists and others turn to that person or group for follow-up information." Finally, notes the moderator, the discovery of any such cover-ups costs government the trust and confidence of its citizens and workers.[33]

And then there is that "dusty desk" discovered at top-security USAMRIID in Fort Detrick—the one that turned out to be laden with anthrax. Researcher Bruce Ivins discovered the anthrax leak in December 2001 during the investigation of the Amerithrax mailings—the same Bruce Ivins who would later apparently kill himself while under FBI investigation as the mailer. However, Ivins initially kept the discovery to himself. He later said in a sworn statement, "I didn't keep records or verify the cultures because I was concerned that records might be obtained under the Freedom of Information Act." He added that he believed the problem solved after he scrubbed the desk and area with bleach.[34]

In fact, the statement was part of an Army report obtained by the *Los Angeles Times* under the Freedom of Information Act. That report also showed a loss of trust in the institute by some of its researchers. "The safety program may be more about insulating the Institute from criticism than protecting the workers," a lab supervisor was quoted in the *Times* as telling an investigator. "Other workers have mentioned that they might not report [lapses] in the future because of fallout from this episode.... I think there's a serious problem."[35]

Some lab errors are not directly caused by humans. The Madison aerosol chamber is touted by its creator as foolproof and leakproof. Three researchers using the device in Seattle were infected with tuberculosis when the chamber leaked, though at least they immediately reported the incident.

Not so at Texas A&M University. Researchers were training on the Madison chamber, using it to infect mice with the pathogen *Brucella.* One researcher came down with brucellosis and was home sick with it for weeks before her doctor diagnosed the condition. She caught the disease when she climbed into the Madison chamber in order to clean and disinfect it. Although she was presumably suited up, *Brucella* apparently got in through her eyes. A&M officials did not report the incident until it had been uncovered by the Sunshine Project using the state's public information act.[36]

"They looked the law in the face and they ignored it," Sunshine Project director Hammond told a reporter. "It's not a lack of training, it's not a lack of knowledge, it wasn't ignorance. It's apparent it was a very deliberate decision.... Nobody learns from mistakes that are covered up."[37]

Subsequently, the Sunshine Project found that Texas A&M failed to report yet another accident in which three researchers tested positive to exposure to the bioweapons agent Q fever in April

2006, although none became ill.[38] The CDC ordered Texas A&M to cease all select agent research until it can complete an investigation.[39]

Some of these accidents occurred among some of America's besttrained scientists and technicians. But according to the Sunshine Project, 97 percent of the principal investigators who received key agency grants from 2001–2005 to study many of the pathogens we have discussed "are newcomers to such research."[40]

All the experiments we've been discussing will be pushed through a collection of hundreds of laboratories scattered through densely populated cities and suburban areas alike, conducted by a population of scientists, students, and technicians with varying skills, training, and social adjustment. If the past is prologue—and when has it not been?—mistakes will be made by some of these fifteen thousand people.

For example, scientists at the Southern Research Institute in Frederick, Maryland, sent dead anthrax bacteria to researchers in Oakland, California—or so they thought. The West Coast investigators injected lab mice with the strain hoping to produce antibodies against anthrax, which they believed might lead to better vaccines. As in Australia, the mice died, but this time the cause was simple. The anthrax was live.

In an eerie 2009 near-repeat of that mistake, Baxter International sent to a lab in the Czech Republic samples of the yearly flu virus accidentally contaminated with the deadly H5N1 bird flu virus. The mistake was discovered when ferrets died when injected with a vaccine made from the samples.[41]

And of course there are those three plague-infected mice that disappeared from a BSL-*3* lab in Newark, New Jersey. Wherever they may be—alive, dead, or the fiction of an accounting error—here's all we know: the FBI crawled all over the lab, intensely questioning animal care workers, researchers, and others, some of whom may have had visions of Guantanamo during the ordeal. No answer was found. The neighboring community, understandably, was left frightened and angry.

The critical point remains that this accident was reported, unlike some others, yet the FBI still never found the answer despite well-publicized efforts and community outrage.

According to the Newark *Star-Ledger,* a federal official said the mice may never be accounted for because they may have been stolen, eaten by other lab animals, or misplaced in an accounting error. The state's health commissioner said that if the mice had escaped the disease would have killed them by the time the investigation began.

The health commissioner's statement was anything but reassuring, implying that the strain indeed was deadly since the mice "would have already died from the disease" by the time he spoke. What if other mice, or fleas—the carrier of bubonic plague to humans—came in contact with the missing mice?

Four years later, déjà vu:

DEAD LAB MICE LOST FROM UMDNJ FACILITY
by Ted Sherman and Josh Margolin
Friday February 06, 2009

The frozen remains of two lab mice infected with deadly strains of plague were lost at a bio-terror research facility at the University of Medicine and Dentistry of New Jersey in Newark—the same high security lab where three infected mice went missing four years ago.

The latest incident, which led to an FBI investigation, occurred in December but was never disclosed to the public. University officials said there was no health threat.[42]

Why no health threat? Certainly an accounting error, officials said, and unlike four years earlier, they did not disclose the "missing mice" because they did not want to alarm anyone. Let's accept that this was "merely" an accounting error. Precise accounting is an important means by which lab workers and neighbors can be assured of their safety from dangerously infected animals. And as we will see, such mistakes do occur, and they are usually covered up to avoid public criticism and scrutiny—not to avoid alarming anyone.

Accidents abound. The 1957 "Asian flu" was no 1918-level killer, but it did take the lives of four million people around the world, many in the United States. Nevertheless, recently, *five thousand laboratories* in eighteen countries were sent samples of the virus to use for quality control tests, presumably by accident. Benign virus samples should have been sent instead. The World Health Organization urged all the labs that got the virus to destroy it to avoid the risk of a new pandemic, however small.

What about escapes from the impeccably designed BSL-*4* laboratories that harbor nature's deadliest microbes? When SARS escaped in Taiwan, through infection of a lab worker, it was from a BSL-*4* laboratory.[43] Fortunately, the Taiwanese lab worker did not infect anyone else. And the recent escape of foot-and-mouth disease virus in England was from the Pirbright BSL-*4* laboratory.[44]

Foot-and-mouth disease results from a virus that infects clovenhoofed animals such as sheep and cattle, but not humans. According to the government organization responsible for health and safety regulation in Great Britain, the Pirbright escape was likely due to problems with the liquid waste disposal system at the lab. The Health and Safety Commission found evidence of cracked pipes and unsealed manholes that caused long-term damage to the site.[45]

The escape of the virus was discovered when cattle became ill at two farms a few miles from the Pirbright lab. Analysis of the virus from infected cattle revealed that the strain, designated 01 BFS, was researched at Pirbright, pointing to the lab as the source. Foot-and-mouth disease does not occur often in Great Britain. The commissioners believe construction at the lab allowed the virus to escape into the soil, which was churned up and carried off by trucks.[46]

Biological Weapons against Agriculture

In a far worse foot-and-mouth disease outbreak in Great Britain in 2001, more than ten million sheep and cattle were slaughtered to stop the epidemic, with costs estimates varying widely from $16 billion to $48 billion.[47] The fact that the infections at Pirbright were quickly contained probably saved even more billions of dollars, a very high cost for a biosecurity lapse. As with SARS, the world dodged another bullet.

Since foot-and-mouth disease does not infect humans, how might it function as a bioweapon? By ruining a nation's economy. The economic importance of crops and cattle led past bioweapons programs to create agents that would attack agriculture. In developed countries, such weapons certainly would damage the economy, perhaps crippling agricultural sales and exports. But given their diversity in agricultural stocks and food surpluses, famine-level shortages would be unlikely. In developing countries, however, it would be a different story. Damage to a major food crop could well lead to massive starvation.

Adding to the seriousness of such threats, plant and animal pathogens may be easier to steal and more likely to escape than those that attack humans, because containment and security might be more lax—a fact that would appeal to terrorists. However, biosecurity experts disagree whether terrorists would be likely to launch such an attack.

Veterinary epidemiologist Martin Hugh-Jones argues, "The British [foot-and-mouth disease] epidemic has given a blueprint to any terrorist."[48] But others counter that terrorists want to instill fear by killing people, so they wouldn't be interested in attacking agriculture. "People attracted to terrorism wouldn't be as attracted to this," says Rocco Casagrande,[49] an expert on agricultural bioweapons. And some plant scientists also think a successful attack on plant agriculture would be unlikely because infecting plants is difficult.

Ominous Opportunities

Lurking in laboratory refrigerators around the world are culture plates and test tubes of all kinds of microbes, many long forgotten and bearing indecipherable labels. The researchers who placed them there have moved on. A major fear is that among all the harmless denizens of the microbial world are contagious strains of pathogens that might accidentally be unleashed. These, at least, are nature's own.

At the BioThreat Characterization Center's home in Fort Detrick, scientists wanted to study ways to make the most dangerous microbes more so. Chief scientist Bernard Courtney told the *Baltimore Sun* that one of the BTCC jobs would be "to create pathogens to match strains that terrorists are clandestinely producing and then develop vaccines and drugs to combat them,"[50] and we should all be relieved that government officials now say they won't take that path.[51] But what were they thinking at the time? Who were the terrorists with the know-how and resources to carry out such brilliant work? They were—and are—figments of someone's overactive imagination, and there is no evidence that that mindset has changed.

Bioweapons control expert Jonathan Tucker noted the Bush administration contended that science-based threat assessment is needed to bridge the gap between the emergence of new bioterror threats and the development of countermeasures. Tucker then gave four reasons why that rationale is flawed:

"First, the administration's biodefense research agenda credits terrorists with having cutting-edge technological capabilities that they do not currently possess nor are likely to acquire any time soon.... Second, prospective threat assessment studies involving the creation of hypothetical pathogens are of limited value because of the difficulty of correctly predicting technological innovations by states or terrorist organizations.... Third, by blurring the already hazy line between offensive and defense biological R&D, science-based threat assessment raises suspicions about U.S. compliance with the BWC and fosters a 'biological security dilemma' that could lead to a new biological arms race." And finally, "the novel pathogens and related know-how generated by threat assessment work could be stolen or diverted for malicious purposes, exacerbating the threat of bioterrorism."[52]

Billion-Dollar Questions

The Center for Arms Control and Non-Proliferation has identified more than a dozen bioactivities that are dangerous. Some activities are justified, some are not. The ones we have discussed are among the most dangerous, and most are really not justified because the risk outweighs any possible benefit.

Dangerous activities spawn question after question. How dangerous are they? Are lab workers adequately protected? What about the rest of us? Is the right level of biocontainment being used? Are there preventatives or cures in case of an escape? Do benefits outweigh risks? Are there safer ways to do this experiment?

Only scientists expert in the molecular biology of pathogens are in a position to have a meaningful discussion of the particulars of a specific experiment. Answers to these questions and a decision to move forward or stop must be considered on a case-by-case basis. Many activities are justified because those questions are affirmatively answered.

Now we want to make an entirely different point, and that concerns not single experiments but the faulty process in approving all of them. Here is a truth that may surprise you if you are not involved in such research: knowledge of a dangerous bioactivity hardly ever reaches the scientific community, let alone the public, before it is underway. Sometimes we don't even know of it until after it is completed. These are no times for surprises, and yet scientists are constantly surprised.

We must have mandatory intensive oversight of dangerous lab activities before they begin. Decisions cannot be left to the scientists conducting the research and their institutional biosafety committees alone. Not all scientists act with the best of motives. They, too, can be swayed by ambition or financial prospects. They, too, are subject to such human failings as underestimating risk and being lulled by a false sense of security into making mistakes and having accidents.

Laboratory work is complex. Mistakes happen all the time.

Both uninvolved scientists and the public must weigh in before anyone is allowed to go ahead with dangerous work that puts us at risk. We should not be dazzled by successes in the lab used as the justification of such work. When researchers find something new, they can say "See! We're learning something," implying that their decision to press forward in dangerous research is justified. This seemed like a fine rationale when researchers discovered that the 1918 flu caused a massive immune response in victims' lungs, explaining why so many had died. But since the bug didn't escape, those opposing the work can't respond with "we told you so." Unless it does escape. If something that deadly should escape into the population, those of us sounding the warnings will have won the day, but at a cost neither we, science, government nor, especially, the public wants to see paid.

CHAPTER SEVEN

Who's Minding the Store?

The story of government control of recombinant DNA, dual-use research, and dangerous experiments can be summarized in one word's dual meanings.[1]

o • ver • sight

1. an unintentional failure to notice or do something: ...[W]as the mistake due to oversight?
2. the action of overseeing something: effective oversight of the financial reporting process.

Who provides oversight for the sprawling efforts to conduct experiments at the appropriate biosafety level? That largely would be the National Institutes of Health, the granting agency for most experiments going on in a few thousand university laboratories. Down at the local level, we generally trust research institutions' legally mandated institutional biosafety committees (IBCs) to safeguard our neighborhoods from harm by providing expert oversight under the aegis of the NIH.

Yet the stories we have been relating of the research in university and government laboratories are so replete with errors and unintended consequences it might seem that in fact no one is minding the store. From the unexplained accidents to the dangerous experiments unchecked—frequently and with no apparent consequences—we seem to be living in a world of oversight in its first meaning, with almost none of the hoped-for second.

But it is not quite that simple. It is not lack of oversight, per se, but built-in failure at critical junctures, going all the way back to the birth of genetic engineering with its federal recombinant advisory committees in the 1970s, that has brought us to this pass. The effect has been to twist the seemingly solid principles and safeguards for recombinant DNA into the surreal failures we have seen.

At another extreme, overzealous elements of the Select Agent Rules enacted after 9/11 have often impeded research and been used to persecute some scientists. Are they, at least, functioning well except for those aberrations?

Let's consider the rules governing select agent use. The *Public Health Security and Bioterrorism Preparedness and Response Act of 2002*[2] and the Select Agent Rules [3] mandate a series of oversight procedures that seem reasonable for those handling the eighty or so agents on the list. The act requires the Departments of Health and Human Services and Agriculture to take the following measures:

- Regulate safety training and physical security in all laboratories that handle select agents.
- Register both the laboratories and any individuals possessing and using the agents.

- Oversee the inventorying of select agents, including assembling a database of any characteristics that would allow particular strains or isolates to be "fingerprinted" as coming from one particular laboratory.
- Join with the attorney general to create a system for identifying those who would be classified as restricted persons under the U.S.A. Patriot
- Act, or who might be "reasonably suspected" by a federal agency of being involved with any organization planning terrorism or other violent actions, or of being an agent of a foreign power, and to deny them access to select agents.
- Inspect registered laboratories to make sure they have adequate physical safeguards in place to prevent select agents from being lost, stolen, accidentally released, or accessed by anyone unregistered.

Penalties for violating these regulations are stiff, ranging from fines of $250,000 for individuals and $500,000 for institutions to imprisonment for up to five years.

Given the history of slip-ups just described, this sounds more like what we need: impressive arrays of regulations and punishing deterrents for those not following them. But the Select Agent Rules deal with security, not accidents or dangerous experiments. Are they at least working as security rules? Apparently not—even within the government's own agencies.

The Department of Agriculture audited its Animal and Plant Health Inspection Service, comparing the regulations against APHIS's actions, and found a string of violations: failing to properly restrict access to select agents and toxins, failing to correctly train those who were authorized to have, use, or transfer them, failing to maintain up-to-date and accurate inventories, and more.

These were no mere technical violations. The audit concluded that the Select Agents and toxins were left "vulnerable to potential theft or misuse."[4]

Meanwhile, HHS reviewed fifteen universities for compliance between November 2003 and November 2004 and found that, although all fifteen had appointed someone responsible for oversight of their select agent work, eleven had weaknesses "that could have compromised the ability to safeguard Select Agents from accidental or intentional loss."[5]

The root of these failures probably lies in the free-spirit culture of scientists unaccustomed to regulations and suspicious of them, and the inability of the already-dysfunctional IBCs to deal with the new era of security regulations.

In the summer of 2008, an update to the Select Agent Rules was introduced in both the House and Senate. The update died but was reintroduced in the House in February 2009, with trivial changes to the earlier version and a new title. The Program and Biosafety Improvement Act of 2009 [6] in fact calls for many improvements, although, unfortunately, it offers no major changes to those Select Agent Rules.

The proposals in the bill we hope to see pass would:

- improve oversight of high biocontainment laboratories, evaluate the need for more of them, and improve training for laboratory personnel;[7]
- create "an integrated Biological Laboratory Incident Reporting System" for lab workers to "voluntarily report biosafety or biosecurity incidents of concern";[8] and
- administer this much-improved reporting system through an independent contractor, "a public or private entity *that does not regulate biological laboratories.*"[9]

On the down side, the existing Select Agent Rules would simply be reauthorized with few major changes. The transparency of activities in BSL3 and BSL4 labs would not be improved, as we believe it must be.

Significant changes in the Select Agent Rules are sorely needed to remove impediments to research. An important change we strongly recommend would reduce the number of select agents from the current eightytwo to perhaps ten of the most dangerous. Most of those eighty-two are not very good candidates for bioweapons; on top of that, they have been researched safely for years without a security hitch. It's unlikely that terrorists would be interested in any but a handful—or could use them to great effect if they tried.

But some change may be in the wind. The bill also calls for the relevant agencies to contract with the National Academy of Sciences to review the Select Agent program, focusing on "the extent to which the program has enhanced biosecurity and biosafety in the United States" and the existing program's effects on international collaborations and scientific advances at home.[10]

The Institutional Biosafety Committees were created as genetic engineering research got its start in the 1970s, and they are required to be open and transparent in a variety of ways, such as having two community members on board who were not affiliated with the institution, and making their minutes and submitted documents public. But a few loopholes lay nested in the legislation that set up the IBCs for failure. First, institutions can pick their two community members—and they have little motive for choosing watchdogs. They can strike a variety of sensitive materials from their minutes and documents—judged sensitive, of course, by the IBCs themselves. Finally, most of the law's oversight provisions are guidelines and not legally enforceable. The latter might not have doomed the groups' work to failure, because the NIH can withhold funding from those violating the guidelines. But the agency doesn't and won't: too much vital research might be impeded. Even prestigious universities pay only lip service to the guidelines, many not even that.

Of course, the vital details that would yield a pointillist image of how biological research is regulated in the real world lie in the IBC reports themselves, those warehouses of documents required to be filed in compliance by all the committees that have met over the years, which sometimes actually are filed.

We need to zero in on the details harboring the devils, and they are legion. Fortunately we can find many without mining years' worth of IBC meetings, because another investigator has already done so. Under the NIH guidelines, minutes of IBC meetings "shall be made available to the public upon request." Following that rule, Ed Hammond of the Sunshine Project filed hundreds of Freedom Of Information Act requests, yielding a treasure trove of reports that are frequently startling in documenting potential hazards passed over without even perfunctory review. Perhaps more important, they reveal how a well-intentioned system veered so dramatically off course.

Of 438 institutions from which Hammond requested minutes, he got them from only 291. Of those 291, many were so heavily redacted there was little information left; in other cases, he was told no meetings had taken place, despite their being required. Hammond noted, "NIH continues to allow these labs to operate without, as far as we can tell, even admonishing them."[11]

But two sets of minutes from North Carolina institutions provided a veritable gold mine—although certainly not in ways to comfort those relying on IBCs for safeguards in potentially dangerous experimentation. Both sets of minutes were heavily blacked out, presumably to keep critical information from reaching the wrong hands.

The assistant general counsel to the University of North Carolina at Chapel Hill explained the heavy redaction of minutes this way: "What kind of worried me about the request was that some of the people on the [Sunshine Project] board were from foreign countries where there had been terrorist cells found, or where I think I remember some assertion by the feds that some rebel group in the country was allied with Al-Quaeda *[sic]*."[12]

The attorney continued that UNC's committee had "deleted the names of a few bugs/toxins that the researchers were working on—they weren't necessarily Select Agents but some stuff we'd just as soon terrorists didn't know we had around."[13]

It is worth pointing out that when information is made *public,* it hardly matters who is on the requester's board of directors—in this case individuals from Asia and Africa, along with staff members from Colombia. In addition, however, Hammond notes that the deleted information can be found in the university's faculty publications and NIH records, as well as on its own Web site.

But, you might ask, how did Hammond get the attorney's letter, which was sent to a colleague at sister Eastern Carolina University? It was accidentally included in the package of blacked-out materials from ECU. But the tale gets much better.

Hammond discovered that the ink used to redact paragraph after paragraph of IBC minutes was sufficiently translucent for him to read the blacked-out material quite easily. Here are some examples of what ECU deemed necessary to keep from public eyes:

"Please refer to Addendum to Biological Safety Policy in the Agenda Packet."

"No glass will be used in the lab."

"Conclusions: The registration was passed."

And the pièce de résistance: "A keyboard cover will need to be installed on the computer."

However, not all the editing was so reminiscent of Yossarian's darkly humorous expurgations in *Catch 22.* The IBC completely blotted out a paragraph discussing the university's malfunctioning waste incinerator, which Hammond finds "more likely an attempt to avoid embarrassment and/or uncomfortable compliance questions about ECU's attempt to improvise a solution for the serious design flaw in its equipment."[14]

In three years of IBC minutes, Hammond found only one mention of select agent research. That was for work on *Brucella,* yet references to it were blotted out of the minutes, in spite of its being documented in NIH public records and on ECU's Web site. However, in this case, Hammond suggests that the committee may have had a motive for trying to hide such a well-known project involving a select agent.

"There was an accident," he writes. "It came in 2003, when a splashed brucella-containing liquid in an ECU BSL-*3* lab caused an exposure to the bacteria. The minutes provide no further detail, but note that a new brucella exposure policy was adopted. Even this rudimentary information, however, would have been unavailable to the public if the redactions had been made competently."[15]

Funny as the Eastern Carolina situation may be, it would be too easy to dismiss it as a case of inattentive administrators bumbling a serious situation. We suggest it is truly a measure of how unprepared and inexperienced even a major academic institution is in weighing biosecurity demands against its own IBC public disclosure requirements.

But even more illustrative of the failure of the Institutional Biosafety Committees and the supervising NIH to confront the new range of dangers is the University of Georgia's research project with none other than the 1918 flu virus.

This pandemic flu virus is extremely dangerous, a view shared by both proponents and opponents of its recreation. Several years ago the University of Georgia conducted a series of genetic engineering experiments to bring the virus back to life, as did a few other NIH-funded laboratories across the United States, and it eventually succeeded in doing so. The question here is, how often did the university's IBC discuss the potential risks of these experiments and the ways to safeguard against them? The answer? Never. Remember that this strain killed some forty million people around the world, that virtually no one alive today has immunity to it, and that a major goal in re-creating the virus was to understand its singular virulence.

The Sunshine Project condemned this violation of the guidelines and three years later again asked for the IBC minutes. This time the university complied, in a manner of speaking. The IBC conducted its first recorded meeting ever one week after the Sunshine Project's request, and 20 percent of the members skipped it. The University of Georgia is also conducting research involving the new "scare bug," the H5N1 bird flu, as well as the bioweapons agents anthrax and botulinum, genetically engineered rabies, HIV, and others.

Are other IBCs dealing with deadly risks so cavalierly? Who knows? Does anyone care? Certainly not the Department of Homeland Security, which had named the University of Georgia as a finalist for a contract to build a new BSL-*4*-laboratory, to be named the National Bio-and Agro-Defense Facility.[16] The initial grant will be worth up to $500 million.

But the University of Georgia is in elite company in many ways. Here is a list of organizations in violation of the NIH guidelines.[17]

GOVERNMENT AGENCIES:

Centers for Disease Control and Prevention/National Center for Infectious Diseases
Department of the Army
Walter Reed Army Medical Center

RESEARCH INSTITUTIONS:

Memorial Sloan-Kettering Cancer Center
Scripps Clinic
Indiana University
Rockefeller University
University of Texas MD Anderson Cancer Center

PHARMACEUTICAL COMPANIES:

Genentech, Inc.
Genzyme Corporation
Abbott Laboratories
Bristol-Myers Squibb
GlaxoSmithKline
Merck & Co.

Who is minding the store? By now, the answer should be obvious.

Oversight Lite

The age of the new biology dawned when the powerful tools of recombinant DNA allowed scientists to forge novel life forms, their powers initially limited to the modest redesign of microbes such as bacteria, yeast, and viruses. That newfound power marked the beginning of modern biotechnology with its potential for yielding "dual uses" at every turn, from new medicines to better bioweapons.

Scientists and public agencies were hardly blind to possible dangers. Initially the NIH Recombinant Advisory Committee and mandated local IBCs were seen as solid overseers of recombinant DNA research. In time, however, the obvious safety of most recombinant DNA experiments lulled everyone into the lackadaisical oversight we have just seen. Now the dangers have reached a new order of magnitude with the large increase in dangerous experiments to make use of the new BSL-*3* and BSL-*4* laboratories. And added to those are worries magnified since 9/11 of inadvertently revealing biotechnology information that could aid an enemy in the design and development of bioweapons.

The National Research Council stepped up to the plate in 2003, convening a panel of leading scientists to devise guidelines for research and publication of dual-use experiments. The committee, chaired by Gerald Fink of MIT, met the challenge in a report with the dead-on title of "Biotechnology Research in an Age of Terrorism: Confronting the Dual Use Dilemma."[18] The committee's report established an important new list of research endeavors warranting special oversight, terming these "Experiments of Concern." Included would be:

- experiments that demonstrate how to render a human or animal vaccine ineffective or that would confer resistance to useful antibiotics or antiviral agents;
- those enhancing the virulence of any plant, animal, or human pathogen or rendering a nonpathogen virulent;
- experiments that increase a pathogen's transmissibility or would alter its host range, that is, developing a pathogen infecting only plants that would now infect animals as well;
- any experiments to enable a pathogen to evade diagnosis or detection, or to weaponize a pathogen or toxin.[19]

The committee concluded its job by urging the creation of the National Science Advisory Board for Biosecurity (NSABB) to work out the means for carrying out the committee recommendations for overseeing dual-use and, although this was less explicit, dangerous experiments that are not experiments of concern. For example, experiments on the 1918 pandemic flu are designed to see what made it so deadly, not to make it more dangerous. As a result, they don't fit any of the listed experiments of concern. But the resurrection of that virus in itself was an example of dual-use research. In enemy hands it would be a horrible bioweapon and its genome sequence has already been published.

Among the solid recommendations by the Fink committee were that experiments of concern would "require review and discussion by informed members of the scientific and medical community before they are undertaken."[20] This implies advance oversight by "outsiders," which unfortunately is not what the NSABB is recommending in its draft report on oversight. That board is composed of a distinguished group of scientists and public health and biosecurity experts.[21] Ability and experience are not issues. Conflict of interest is. The indications so far are that the board is avoiding mandating outside oversight, delegating decisions to individual researchers and to IBCs—right where all the problems started. Board Chair Dennis Kasper and several other members have a stake in high-biocontainment laboratories or are involved in experiments that are dual use or arguably dangerous. In other words, the board has a vested interest in little outside oversight.

Shades of Meaning

The ponderous title of the board's draft report on dealing with dual use research oversight hints of its weaknesses: "DRAFT Report of the NSABB Working Group on Oversight Framework Development: Proposed Strategies for Minimizing the Potential Misuse of Life Sciences Research."[22] From that point onward, the focus is on deliberate "misuse" of life sciences research, but neglects the important additional focus of research that is inherently dangerous.

For example, the draft uses a single criterion to define dual-use research of concern: "research that, based on current understanding, can be reasonably anticipated to provide knowledge, products, or technologies that could be *directly misapplied by others* to pose a threat to public health and safety, agriculture, plants, animals, the environment, or materiel."[23] Nowhere is there explicit mention of research that is inherently dangerous, independent of misuse, another indication that the board is protecting its own.

We argue that the second major area needing intensive advance oversight is research that is dangerous in itself, such as manipulation of Asian flu viruses or that on the 1918 flu virus—that is, research that is dangerous independently of its dual-use potential.

The board also obfuscates the clear "concern" in any experiment to hamper the effectiveness of a vaccine, as described by the Fink report, instead limiting concern to experiments disrupting the "effectiveness of an immunization without clinical and/or agricultural justification."[24]

Consider the well-known Australian mousepox experiment. The researchers' intent was to reduce the size of the mouse population that was destroying crops, so it clearly had agricultural justification. However, their findings turned out to pose major dangers. Would the NSABB consider such experiments subject to oversight even if the startling result of the Australian mousepox were known in advance? The Fink report is clear that such an experiment would require *outside* oversight before commencing.

In sharp contrast, the proposed NSABB guidelines would allow only the investigator and perhaps the IBC to make the decision to proceed. And that's not all that is hidden in these details. Another category of concern stated in the NSABB draft report would be experiments to "generate a novel pathogenic agent or toxin, or reconstitute an eradicated or extinct biological agent."[25] Recall that the BioThreat Characterization Center planned to generate novel pathogens so we can defend against them, thus requiring special oversight. Will the BTCC's oversight board—its IBC—deny permission to carry out such experiments? Don't bet on it.

Another key weakness threading through the oversight process from the 1970s comes to life here, too, like a bad gene reasserting itself: these are to be voluntary guidelines. Research on the "eradicated" 1918 pandemic flu virus also would call for special oversight. But we have already seen that the University of Georgia never met to approve or deny permission for experiments reconstituting the virus. Given past performance, why would anyone believe that a "call for" special oversight would be heeded by anyone under yet another set of voluntary guidelines?

Finally, when the report turns to the public's interests, it is entirely confined to improving perceptions. Scientists should communicate with a public "increasingly sensitive to issues pertaining to research involving dangerous microbes and the risk of accidental or intentional release of such agents." Why? Because "a lack of public understanding and appreciation for the reason for conducting ... dual use research, sensationalism of dual use research findings, and concerns about public safety and national security all serve to undermine public trust in the life sciences research enterprise."[26]

In plain English, the report claims that scientists need to placate and pacify an anxious public, not to ensure public safety by enforcing reasonable rules of research conduct.

Even the original NIH guidelines were tiered in an "inverted pyramid," so that more questionable or dangerous research called for increasing degrees of scrutiny. The NSABB report all but abandons even that voluntary caution. Among an institution's responsibilities in dual-use research oversight would be "assisting the [primary investigator] in deciding whether her or his research meets the criterion for dual use research of concern and thus requires further review or oversight." But, "In the great majority of cases, it is anticipated that the institution will rely on the judgment of the researcher."[27]

Ironically, the draft narrows oversight from the tiered process preserved even in the Fink report that gave rise to the board itself.

Biting the Bullet

The NSABB draft, like the Fink report before it, never seriously considers the powerful alternative to the "volunteerism" to which it invariably defaults for compliance. That alternative is regulations that are enforceable and, in fact, are enforced with penalties for noncompliance. Here is the question: Should dual-use and dangerous research be subject to guidelines or enforceable regulations? Many scientists do not want enforceable regulations, viewed as potentially interfering with their ability to carry out independent research. But the increasing numbers of dangerous experiments—where experts disagree over how they can be done safely or even done at all—cry out for international oversight and enforceable regulation. We must gain some control over biotechnology activities.

The exponential growth of biological knowledge and its powerful tools makes controlling dual-use and dangerous research urgent. However, we are not proposing a simplistic set of regulations that would achieve some form of biosecurity at the expense of the enormous potential already demonstrated by the new biology. Without great care, we could end up with the type of draconian, self-defeating regulations that we have seen in some of the Select Agent Rules and their enforcement.

But there are thorny legal issues in moving toward regulations. Jennifer Granick, executive director of the Stanford Law School Center for Internet and Society, anticipated these concerns when she outlined problems that would follow turning the NSABB draft guidelines into law.

"As a lawyer for computer security researchers, it is impossible to regard this prospect with anything but dread," she says. Citing the possibility that the guideline for restricting publication of some dual-use research could someday become law, she says, "the author risks criminal prosecution if law enforcement disagrees with a scientist, university, or peer-review publication's decision that the research should be published.... And, legally, I'd find it extremely difficult to advise the author with any certainty whether publishing the research is lawful or not."[28]

Granick's legal concerns are important, but her solutions lead us back to where we started: "Voluntary self-regulation, ethical training, peer review, and additional practices currently followed by recombinant DNA researchers, microbiologists, and other scientists all have a track record of success. And smart federal laws can control access to pathogens—and prohibit dangerous practices—while steering clear of restricting scientific publications."[29]

We have already seen the countless exceptions to that idealized view of science. However, Granick puts wonderfully concisely the black-andwhite, all-or-nothing choices in which this discussion has always been framed: draconian laws or purely voluntary compliance, as though such complex issues must resolve to binary answers.

Building the Prototype

In a major contribution to the discussion of oversight, the Center for International and Security Studies at the University of Maryland has designed an oversight system that goes way beyond all the proposed guidelines we've discussed. The Biological Research Security System is designed to cover all dangerous experiments in biology—dual use being only a subset—and it is designed to be international in scope, as long-term solutions must be.[30] The fruits and poisons of the new biology can travel as widely and as quickly as a pandemic.

The Biological Research Security System was no cobbled-together brainstorm. The Maryland group developed it over several years with the help of dozens of international experts, ranging from scientists to policy, biosecurity, and arms-control experts to lawyers. This massive undertaking resulted in an extremely well-thought-out and innovative model for future oversight of biology. And the authors acknowledge the many hurdles that must be crossed before a successful, far-reaching system can become operational. For one, they know that "meaningful protection" can only be won by "imposing some constraint on freedom of action ...[on] fundamental research."[31] For the system to work, any constraint's protections will have to justify its costs.

The authors, we believe, propose a system of oversight that solves the two major hurdles posed by dual-use and inherently dangerous research:

- How to appropriately tier the levels and thus the weight of review based on the danger a planned experiment poses.
- How to define that risk objectively.

To see how deftly they accomplish this, let's take the second issue first because it is potentially the most contentious. For example, we would classify projects involving the infective, whole 1918 flu virus as of "extreme concern" because of the potential consequences of escape from the laboratory. The same for altering avian flu genes in the live virus to learn how it might become contagious in humans. But we have already seen that scientific experts debate not only the dangers these experiments pose but even, stunningly, the level of biocontainment required. How can we define experiments of potential, moderate and extreme concern? Here, the authors present a novel yet wholly realistic working definition of "intrinsic danger" as a cross between a pathogen's transmissibility and its virulence,[32] qualities that can be well-defined by epidemiologists. In other words, experiments of extreme concern would involve pathogens that were both highly virulent and highly transmissible figure 7.1).

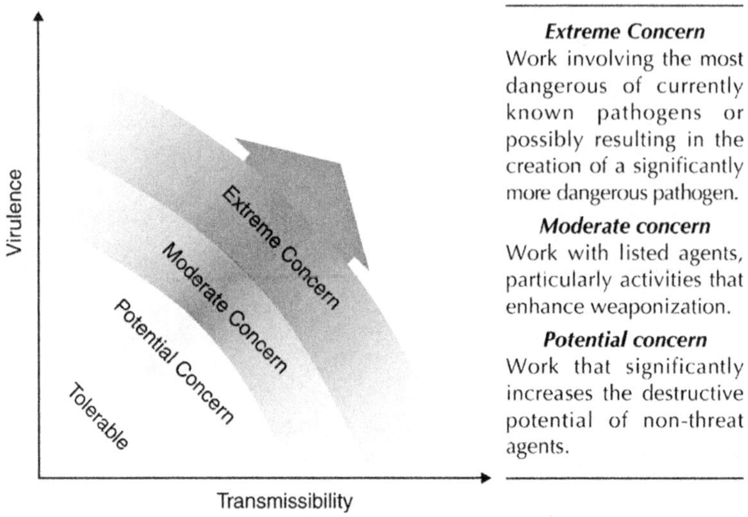

FIGURE 7.1. Conceptualization of experiments of intrinsic danger and their level of oversight (John Steinbruner, Elisa D. Harris, Nancy Gallagher, and Stacy M. Okutani, "Controlling Dangerous Pathogens; A Prototype Protective Oversight System," The Center for International and Security Studies at Maryland, March 2007, p.24).

Now the first issue: How do we determine appropriate level of review based on potential danger? Here is a schematic version of the tiered system the Maryland group has devised, from most to least dangerous:

> International Oversight: *Activities of Extreme Concern*—"An international body would be charged with approving and monitoring all research projects of extreme concern. That authority would be narrowly focused only on those ... that could put an appreciable fraction of the human species at risk, such as work with

smallpox or a yet more lethal contagious pathogen."

National Oversight: *Activities of Moderate Concern*—"National oversight bodies would be responsible for research activity of moderate concern, such as work with anthrax and other agents already identified as having biological weapons potential."

Local Oversight: *Activities of Potential Concern*—This "encompasses those activities that may increase the destructive potential of biological agents that otherwise would not be considered a threat."

No oversight: *All other research.*[33]

The authors say their advanced oversight system features three major innovations. The most consequential research would be subject to international jurisdiction, in our view the only means of ultimately maximizing our biosecurity; oversight would be comprehensive, not hit or miss, for every jurisdiction; and, most significantly, the oversight process would become a legal obligation.[34]

The report stops short of designing a framework for the civil and criminal laws that would have to be devised and passed for the system to work. That would be beyond our expertise as well, but we do see critical elements to be included in that essential framework.

To begin with, at some assessed level of intrinsic danger—and thus "concern" over a biology project—the local IBC might be required to inform a higher-level expert oversight committee, which would decide whether the experiment could go forward. Before that decision, however, the public would have to be fully informed. The exception to this is cases in which the committee judges that national security would be compromised.

Who would be doing those assessments? The government oversight committee should consist of scientists, epidemiologists, public health professionals, and biosecurity and international policy experts. The wide range of expertise is necessary, because decisions could have wideranging impact, all the way from our personal safety to international relations. Finally, the committee should have a majority of nongovernment employees.

Liability would be a major concern for everyone involved in research and oversight under the proposed system. Stanford attorney Granick raised important questions about the potential effects of disputed risk assessments in legal actions. "We know that reasonable scientists can and do disagree about these things," she wrote. "What do prosecutors, judges and juries think?"[35]

Disagreements among scientists are important means of advancing science. Taken into a courtroom, they could lead to legal decisions that drain fortunes or, as under the Select Agent Rules, send a scientist or committee member to prison.

Clearly, individual scientists, IBC committee members, and the government committee members would have to be protected from civil actions, analogously to the way protections are extended to employees of corporations judged with civil liability. The institution would assume a legal responsibility for its researchers and its own overseeing IBC. The institutions themselves would still be open to civil suits and fines, but individuals would not be. In fact, some of these institutions are indeed corporations, so required legal distinctions for liability may already be in force.

Individuals, as now, could be prosecuted for criminal acts committed within the institution. But the definition of criminal action would need to have a high bar and well-defined boundaries, including precise definitions of what constitutes crimes, such as flagrant violation of laws, disregard for physical harm to people or animals, and the murkier concept of "criminal intent." Perhaps all three of these legal barriers would need to be breached for an action to be judged criminal.

For example, the highly regarded plague researcher Thomas Butler would not have been criminally liable under these suggested laws for his errors in judgment, since they entailed no potential harm and he had no criminal intent in the actions he took when he found plague cultures missing.

Placing the civil responsibility with institutions has advantages. It provides another tier of oversight above the researchers and IBCs to make sure they obey regulations or face financial exposure. Meanwhile, researchers and IBC members would be able to do their work, confident that they are immune from fines or civil suits, but well aware of clearly defined criminal acts and their consequences.

Conclusions: Beginnings

Difficult legal and science policy issues will certainly complicate efforts to give oversight regulations the force of law, but they must be resolved. We need to enforce reasonable regulations to protect the public, and we need to protect and not impede scientists in their vital work.

The NSABB offers a false sense of security, applying only to dual-use research and virtually ignoring other dangerous experiments, and leaving all responsibility in the hands of researchers and IBCs. Finally, even if a government committee could provide adequate oversight—as none ever has before—compliance with guidelines or regulations applies only within the United States. Ultimately, no one country can be safe unless oversight protections apply internationally.

CHAPTER EIGHT

All Roads Must Lead to Public Health

To us, biosecurity means safeguarding from infectious disease in all its manifestations. That requires committing the largest portion of our finite resources to shielding against the threats most likely to kill us. Protection against bioweaponry is just one element in such a shield, but in the wake of the Amerithrax attacks and the government's hasty and overblown response, this highly unlikely threat to American life has become the squeaky wheel, garnering billions in appropriations, cornering the market on public fear, and capturing the news spotlight.

We must ask ourselves what biological threats pose the greatest danger to our families. Biowarfare is well down the list of probables, but just as we weighed the likelihood and consequences of various biological attack threats against one another in chapter 4, we can now put some force behind our claims for the low risk of a major anthrax attack, the kind that could truly disrupt public life and would require a major public health response, in comparison with looming health threats such as a pandemic flu and the steady, year-in, year-out toll of annual diseases

FIGURE 8.1. Toles ©2007 Washington Post. Reprinted with permission of Universal Press Syndicate. All rights reserved.

like drug-resistant staph infections and garden-variety influenza.

Now the government assesses the relative likelihood and consequences of bioweapons, pandemic flu, and annual infectious disease threats in three different "boxes" so the yearly threats never get compared to bioweapons. That means the most powerful or emotionally charged concern wins in the ensuing funding battles. Efforts against bioweaponry and pandemic flu are funded from special vaults provided under BioShield 2004, the Bush administration's $6.1 billion pandemic flu plan,[1] and other dedicated sources. That leaves the killers and disablers of the largest number of Americans still wanting for additional research funding. Only a combined risk assessment makes sense to determine a health hazard's true impact.

As earlier, we're using the term "risk" as an indicator of the seriousness of a threat to us, arrived at by multiplying the consequences of the threat by the probability of occurrence. Since our crude assessment considers only fatalities—in order to make the point that our priorities are skewed—some of our conclusions are only a first word on the subject, not the last. Let's take annual flu as the standard against which other threats are measured. Flu kills around thirty-six thousand people *every year,* so deaths in the tens of thousands occur each year with certainty—that is, with a probability of one. Another group of particularly deadly disease agents, the feared hospital-borne multiple drug - resistant (MDR) bacteria, kill tens of thousands every year, again a certainty. A recent report places deaths from just one of these bacteria, methicillin-resistant *Staphylococcus aureus,* at over eighteen thousand per year, an annual death toll greater than that of AIDS.[2]

Let's add into the mix the possibility of a pandemic flu, now a big worry, which is distinct from the annual flu. But how can we rationally assess the threat of a bioweapons attack that has never happened against that of a global flu outbreak that recurs, if infrequently, and annual infectious diseases? The fact is that it must be done for simple reasons. We mount bulwarks against all manner of biological threats—from AIDS, tuberculosis, and other infectious diseases to biowarfare—and we do so from a fund of limited resources that we must spend efficiently.

Without a data-based ballpark calculation of the threat and size of a new pandemic flu, even experts can make wildly varying predictions. Here is the assessment of Robert Webster, not only an eminent virologist but director of the WHO Collaborating Center on the Ecology of Influenza Viruses in Lower Animals and Birds: "The world is teetering on the edge of a pandemic that could kill a large fraction of the human population,"[3] he wrote recently in a coauthored article. Recalling the 1997 virus that spread through Hong Kong poultry markets, Webster said, "the only thing that saved us was the quick thinking of scientists who convinced health authorities to slaughter more than a million domesticated fowl in the city's markets. The avian virus turned out to be a new strain—one that the human population had never seen before. These deadly new strains arise a few times every century, and the next one may arrive any day now." Webster believes the 1918-like scenario will replay. "But this time it will be worse."[4]

Michael Osterholm, director of the Center for Infectious Disease Research and Policy at the University of Minnesota, agrees. He foresees a toll of some 270 million pandemic flu deaths worldwide, a number he arrived at simply by extrapolating the 1918 death toll of 40 million to the world's current population.[5]

Scary stuff, but now listen to Paul Offit of Children's Hospital of Philadelphia and the University of Pennsylvania School of Medicine. He says of avian flu, "the virus is clearly not so highly contagious among mammals, and I just don't think it's going to become so."[6] Moreover, Offit does not expect the next pandemic until sometime around 2025.

The WHO itself takes a conservative position, estimating that a new pandemic would take the lives of from 2 million to 7.4 million people, believing that the 1918 pandemic was unusual and unlikely to be repeated.[7] We will try to make rational risk assessments about pandemic flu and anthrax as best we can in a field of such uncertain numbers.[8] Since anthrax is on everyone's minds, we'll tackle it first. As discussed earlier, a bioweapons assault resulting in thirty thousand fatalities is a reasonable guess for a large attack, so we will use it. What is the probability of an attack of this magnitude in any given year? Is it 001? Or .01? Or a radical 0.1—meaning that there is a one in ten chance of an attack in the coming year? We must make some choice, because everyone's risk assessments are based on guesses of such probabilities.[9] But what does each one mean in everyday terms?

Our intuitive method involves simply thinking about the number of years that must go by to have a 50–50 chance of such an attack, given its probability in any one year. Table 8.1 offers some examples.

To set the probability of such an anthrax attack at 0.1 would mean there is a 50–50 chance that in 6.6 years we will have suffered at least one such attack resulting in 30,000 fatalities. Given what we showed would be required to mount such an offensive, that probability would be unrealistic. Only the most fearful among us would believe this to represent the seriousness of the anthrax threat today. How about the lowest probability, 0.001? We could then expect 692 years to pass before there was a 50–50 chance of at least one attack, and clearly one would have to be blissfully optimistic to pretend we are that safe. Any realistic guess must lie in between, so let's look at 0.01 probability, meaning that 69.9 years will pass before there is a 50–50 chance that an attack has occurred. This, too, seems optimistic, although it seems more realistic, given the 30,000 fatalities that would mark an extremely successful bioweapons attack.

Consider a higher probability: 0.03. Arbitrary, but since we have no way of knowing the real numbers, this seems a reasonable guess. At this probability, we should expect about 23 years to go by to reach even odds that we will have suffered at least one attack. It means that the likelihood-weighted fatalities in any one year would be 900, the product of 0.03 probability and the 30,000 deaths used for the threat assessment. That's more fatalities than the annual toll of tuberculosis but still quite a bit fewer than deaths from multiple drug – resistant staph infection, garden-variety flu, and others. One conclusion is obvious: more funding for such annual infectious diseases could save far more lives. Interestingly, that would remain true even if we had used that unrealistically high probability of 0.1 and extreme fatalities of 150,000. The risk-adjusted fatalities would number 15,000, and the conclusion would not change.

Table 8.1. Likelihood of a bioweapons attack.

Probability of an attack in any one year	Years until 50% probability of at least one attack
0.1	6.6
0.01	69
0.001	692.8

Likelihood is expressed both as the yearly probability of the attack and as the number of years that must go by to have a 50–50 chance of at least one attack.

To arrive at truly realistic probabilities for these attacks would require better worldwide intelligence than we have seen or probably will. We would need to know how close rogue nations are to developing and delivering bioweapons and their plans for using them, as well as whether terrorist groups have the agents and are capable of delivering them to cause mass deaths. On the other hand, if we were to get solid intelligence about plans for a particular attack, the odds estimation would become much more concrete.

Precisely or not, we have to assess threats to plan for countermeasures we want to have in the Strategic National Stockpile [10] ten to fifteen years from now. Decades of experience in new drug development by pharmaceutical companies shows that it can take that long to discover, develop, and produce new countermeasures.

Now consider influenza—not the garden variety but the kind of severe pandemic we know can kill many millions. We estimate that the next pandemic will claim thirty-six million victims worldwide,[11] about the number of deaths from the devastating 1918 pandemic. This death toll is higher than the conservative WHO estimate but far lower than Osterholm's, so it strikes a balance. There is also some statistical justification for using this particular number, which is the logarithmic average of the WHO's and Osterholm's projections.

The U.S. population of 300 million is about 4.6 percent of the world's 6.5 billion, so we will estimate the number of U.S. deaths at 1.7 million—although America probably would not suffer as great a loss of life as developing countries.

For the chances that such a devastating outbreak will occur in any one year, some real data are available from the actual influenza pandemics of 1918, 1957, and 1968,[12] and if the next pandemic had started as we did this projection (2007) it would be recurring an average of every 29.7 years.[13] Granted that a total of three pandemics in a century provides meager data, but they are certainly better than the data for anthrax, which are nonexistent. So we will estimate the probability of such a flu pandemic in any year as one in 29.7, or 1/29.7=0.034 yearly probability.[14] That would put the likelihood weighted yearly U.S. fatalities at 57,200—representing 0.034 times 1.7 million. That is more than the yearly toll of any single infectious disease and about a third of the 177,000 fatalities from all infectious diseases in the United States.[15] The bottom line is that the government is clearly on the right track in committing large-scale funding to research and development of pandemic flu countermeasures. However, there are concerns over how those funds are being used. Too much money has been spent on Tamiflu, a drug that may not work on pandemic flu, and as we discuss later, more money is urgently needed at state and local levels to improve sorely needed surge capacity at hospitals.

Some argue that a massive bioweapons attack would heavily damage our national security, so we must counter even the slimmest of possibilities with concomitantly massive funding. In this view, virtually any amount of spending on biodefense is justified, and comparisons with budgets for annual infectious diseases are irrelevant.

Those who justify such biodefense budgets further argue that Congress, which appropriated the massive amounts for bioweapons countermeasures, has judged such funding levels appropriate and is simply carrying out the will of the people.

On strategic grounds, Gerald Epstein points out, "A nation's security infrastructure exists to protect its citizens' lives, but that has never been its sole responsibility. More generally, it exists to preserve objectives that include national sovereignty and freedom of action." He further adds, "Policies that preserved the lives of every American citizen but ceded control over U.S. foreign policy to others would be rejected by any American political leader."[16]

We agree. But deflecting badly needed new funding away from infectious diseases with their high yearly morbidity and mortality and from such growing threats as antibiotic-resistant bacteria in order to bolster bioweapons defense also represents an obvious national security cost. Victor Sidel sums up: "In short, bioterrorism preparedness programs have been a disaster for public health."[17]

As we have shown, our massive biodefense-focused spending is out of proportion to the current threat. Of equal importance, much of it is misdirected. Development of countermeasures to natural infectious disease would offer direct benefits for biodefense efforts—far greater than those that, conversely, would accrue to infectious diseases prevention from biodefense research. We *can* have our cake and eat it too—by funding research and countermeasure development for natural infectious diseases and keeping a sharp eye out for the many potential biodefense spinoffs.

Yes, by its votes Congress has demonstrated that it believes massive biodefense funding is the way to go, but we think the majority of members and their constituents were swept up in the Bush administration's fear campaign, so the very real threat of political fallout is inhibiting good decision-making.

But help may be on the way. Alerted by a host of recent high-profile accidents, the House Energy and Commerce Committee Subcommittee on Oversight and Investigations held a hearing entitled, "Germs, Viruses, and Secrets: The Silent Proliferation of Bio-Laboratories in the United States."[18] The Democratic subcommittee members grilled CDC and NIH testifiers on the need for such a large number of labs as well as on safety in the labs. Keith Rhodes of the General Accountability Office, the government's own watchdog, had this to say regarding the BSL-*3* and-4 labs working on select agents: "The fact that there is so much unknown at the moment, I would have to say there is a greater risk to the public."[19] In the broader context of U.S. biosecurity policy, that "greater risk" is the major theme of this book. Congress is following up. The Senate and House bills updating the Select Agent Rules [20] call for the National Academy of Sciences to conduct a review of the select agent program to understand its effects on international scientific collaboration and scientific advances in the United States. A unified bill probably will be enacted in 2009.[21]

But perhaps the strongest argument we can make is one we introduced in other contexts: our massive and mostly secret biodefense program is creating suspicions among other nations, and those suspicions are damaging whatever security we do have.

In Case of National Emergency, Do Not Complain

Here is how the government plans to respond following declared emergencies that affect our national security, emergencies ranging all the way from lethal pandemics to bioweapons attacks. Because we all want to be vaccinated or treated as quickly as possible, if no proven treatments are available from the Strategic National Stockpile, the government may decree that untested remedies can be used on everyone.[22] There is some indeterminate chance they could maim or kill. And there is more. If any number of people have injurious or fatal reactions to any of the measures taken in this emergency, neither they nor their survivors can claim damages.[23]

The government has indemnified the drug companies that developed the untested remedies, because otherwise they would never have allowed their use. If that sounds like a cure worse than the disease, it is sum and substance of current plans in the event of such emergencies. Emergency use of untested remedies is a part of the tacked-together act that governs all the issues we've been talking about: BioShield 2004. A newer act, Bioshield 2006,[24] is more rational, but the 2004 emergency provision was not nullified, so it remains in force.

Being inoculated with an unproven vaccine or taking an unproven drug in the name of a "shield" is bad enough, but how could Congress leave the country with no recourse for compensation in the face of harm? The last element was literally sneaked into law in the dead of night. According to Public Citizen Congress Watch, "The measure, which was never considered in committee, was surreptitiously set up for passage under cover of darkness. The then Senate majority leader Bill Frist (R-TN) inserted the 40 page text into the Department of Defense appropriations bill late on a Sunday night."[25] A House and Senate conference committee had already completed work on the bill, and the report says that "those conferees had already received assurances from the leadership that the controversial liability shield would not be included in the spending bill."

Here is a preview of how such an emergency declaration might actually play out, as it did in microcosm in 2004. Paul Wolfowitz, then deputy defense secretary, called for the anthrax vaccination of all U.S. forces in South Korea, the Horn of Africa, and the entire Middle East. The personnel were not informed of the risks and side effects, and consent is not required from those in the military. Wolfowitz said that based on "a classified November 2004 intelligence community assessment, I have determined there is a significant potential for military emergencies involving heightened risk to U.S. military forces of attack with anthrax."[26]

The CDC claims that serious allergic reactions to this vaccine are very rare, occurring in fewer than one in a hundred thousand vaccinations.[27] But the

National Vaccine Information Center sharply disagrees.[28] The nonprofit educational organization says, "Since the 1991 Gulf War, healthy young soldiers have reported severe deterioration in health following anthrax vaccination, including chronic joint and muscle pain, weakness, gastrointestinal disorders, disabling fatigue, loss of memory and ability to concentrate, severe headaches, respiratory and heart problems, and autoimmune disorders that leaves them unable to work or live a normal life." The report added that the FDA had received five thousand adversereaction reports demonstrating that anthrax vaccine is causing serious health problems, "but still issued a final order declaring the vaccine safe and effective."[29]

The NVIC is not alone. Mark Geier, an expert on anthrax vaccine and its complications, summarized data from the *government's own adversereaction reporting system.* It showed that there have been 825 adverse reaction reports per million vaccinations for anthrax vaccine, compared with only 34 for tetanus vaccine; 15 disabilities per million for the anthrax vaccine versus 1.1 for the tetanus; and 1.2 deaths per million, compared to 0.13 for tetanus.[30] We don't know where the CDC got its data, but its estimate would predict only 10 adverse reactions per million shots versus the 825 Geier culled from government data.

Geier concluded, "Anthrax vaccine is associated with a series of serious adverse events that can significantly impact multiple organ systems within the body, and result in permanent disability."[31]

The U.S. military now has plans to vaccinate as many as 300,000 personnel.[32] Using Geier's numbers, that might cause 250 adverse reactions, including 5 permanent disabilities. That may seem like a small number especially given the higher risk acceptance for the military, but there is no evidence that any of our enemies in the Middle East possess virulent anthrax or the means to weaponize or deliver it. It is even more unlikely that they have developed antibiotic-resistant anthrax. Wouldn't it just make better sense to make antibiotics available throughout the Middle East and spend the enormous cost of an anthrax vaccination program on real lifesaving protections such as body armor and heavily armored vehicles?

There are similar issues with smallpox vaccine. In an article pointedly titled, "Risks of Being Risk-Averse," Paul Offit gives a relevant example:

> In March 2003, when the United States invaded Iraq, the Department of Defense feared a biological counterattack with smallpox. So it inoculated soldiers with smallpox vaccine; 40,000 healthcare workers also were immunized. Since December 2002, about 1.2 million people have been immunized.... Unlike the prevention of many infectious diseases, smallpox vaccine works even when it is given 48 hours after exposure to someone with smallpox, a disease whose symp-

> toms aren't subtle. We could have distributed the vaccine, made sure that systems were in place to give it quickly and efficiently, and waited.... But we didn't. As a consequence, about 140 people were harmed when the vaccine virus caused inflammation of their heart muscle. In this case, getting vaccine was riskier than waiting.[33]

What were Frist and other supporters thinking of when they put everyone in this potentially precarious position? First, that there would be no way to get the major pharmaceutical firms rolling on such a massive project as the Strategic National Stockpile without such protection—and for good reason. The companies are well aware that if they were to meet the government's order for such untested drugs and vaccines, then had to fight lawsuits for adverse consequences, they might well face bankruptcy. The fatal flaw lies "upstream," in the rush of potentially dangerous countermeasures into massive public vaccination and treatment programs.

What, then, should Congress do? If such a pandemic were to break out, could we wait years while potential lifesavers went through the tedious but safer system of clinical trials? The 1918 flu did its lethal work in little more than a year.

There is no simple solution to this problem, but we will offer some ways to get things moving.

First, some background on clinical trials. It would take us too far afield to run through the intricate process by which drugs that are candidates for market in the United States go through the three phases of clinical trials required for FDA approval, but there are many excellent articles and Web sites that feature this information.[34] Remember that any drug that has been pulled off the market because of significant adverse reactions has completed all of its clinical trials. But often it is only when a new treatment is applied to a large population in need of it that these reactions begin to show up. Yet new drugs are not brought to market in haste—quite the contrary.

It typically takes eight years of clinical trials before the FDA approves a new drug for market. Not only is that an eternity if the nation is trying to prepare for a pandemic, bioweapons attack, or other major health emergency, but only one in five drugs entering phase I trials to test for safety ever makes it all the way through.[35] Those are not good odds in the face of potential disaster.

Clinical trial phases II and III test for efficacy—but still test for safety as well. Of the vast majority of drugs that don't make it, one in five fails for safety reasons and nearly half for efficacy, even though most of them were tested in animals before entering human trials. The rest are abandoned for business reasons. Learning that a drug has proven safe and effective in animals still leaves major questions about what it will do in humans. Whatever the solution to the time-lag dilemma may be, allowing countermeasures to be used that have never been tested in people should not be part of it, except when nothing else is available and the emergency is extreme.

One Possible Solution

The safety problems we have raised regarding untested drugs might be solved by a "Safety Clinical Trial Rapid Response Network," which we now propose. The Rapid Response Network would be set to begin drug safety trials the moment a countermeasure reaches the point of human testing, and the new drug would move quickly through a series of safety trials. Only if successful would it be included in the Strategic National Stockpile. This would reduce the gap between the end of animal trials and adequate safety trials in people.

Here is how it might work. From a pool of patients recruited well in advance, safety trials on fifty subjects, similar to phase I trials, could be carried out in a few months with expedited FDA review. At this point, there would be some confidence in the countermeasure's safety, but safety testing would not stop here. A second trial with two hundred waiting subjects—about the same number as in current phase II trials—would be carried out over the next few months. Then a third trial would be conducted with a few thousand subjects over the next year, as in phase III trials. At this point we could be confident that the countermeasure was at least as safe as a newly marketed drug tested under the current system, yet we would be getting drugs into the Strategic National Stockpile in about two years.

That overcomes the safety hurdle, but it still fills the stockpile with drugs that might be useless in an emergency, because they've never been tested for efficacy. We can't infect people with anthrax, tularemia, Ebola, or the agent of another potentially lethal disease to see if the cure works. But we don't need to.

Those patients already exist, often in large numbers—just not in the United States. Most of these agents occur naturally somewhere in the world and often throughout developing countries. Bringing their victims into government-sponsored clinical trials using drugs already tested safe might cure them of an otherwise disabling or fatal illness that has no other treatment, and at worst it would do them no harm. That would serve the cause of international public health and increase our goodwill. Setting up such a program should be a goal of future legislation.

Shouldering Blame

We still face the second problem. Those treated must have recourse from adverse reactions, but at a cost that will not drive drug companies out of the countermeasure field, and vaccine companies in particular have a long history of facing such liabilities—even electing to shut down operations because of them. That brings us to the Salk polio vaccine, one of the twentieth century's "miracle drugs."

It was in the heady days of 1955 when parents were suddenly able to free their children and themselves from the annual terror of the summer polio season. Jonas Salk had created a vaccine using live polio virus that had been inactivated by formaldehyde.[36] Now the immune systems of those vaccinated would encounter the virus and mount a permanent defense against it, but the so-called inactivated virus could not hurt them. That was how it was supposed to work and nearly always did—except for "adverse events."

In the vaccine made by Cutter Laboratories, one of five manufacturers, some virus remained activated. The results, mostly forgotten after more than half a century, were tragic. Of 120,000 children accidentally given activated poliovirus, 40,000 developed mild polio, but 200 were permanently paralyzed and 10 were killed.

The single lawsuit against Cutter marked a new age in litigation and its often-negative effects on medical practice. A sum of $147,300—miniscule by today's standards—was awarded by an Idaho jury to the parents of an 8-year-old whose legs were paralyzed after she had received the vaccine. "It was a turning point," says Philadelphia's Paul Offit. "Because of the Cutter decision, vaccines became one of the first medical products to be eliminated by lawsuits."[37]

Lawsuits against other drug companies for damages alleged by their vaccines were not always as clear cut, but the effect was unequivocal. In 1950s America, twenty-six companies manufactured various vaccines. By 2004, there were four. Trial juries awarding compensation in staggering amounts often unrelated to the injury were part of the problem.[38] But another was that vaccines have small sales and low profit margins compared to drugs.

To some degree, an answer is already at hand. In 1986, Congress passed the no-fault National Childhood Vaccine Injury Act,[39] which compensates those injured by childhood vaccinations. In a typical year, the program pays out $60 million in claims for damages from childhood vaccines through what is called "vaccine court."[40]

As things now stand, Americans have virtually no way to collect compensation for damages from vaccines used in a declared emergency. This "vaccine court" concept should be quickly adapted for future biosecurity legislation.

The sweeping loss of liability protection in the Frist legislation must be repealed quickly, but repeal must be coordinated with new liability protection for those who develop countermeasures and those who deliver them—usually first responders—that preserves victims' legal rights and provides for compensation to those harmed. Victims harmed by any form of countermeasure would have their damage claims reviewed by experts—research scientists, epidemiologists and clinicians—not juries of peers. In this way, fair compensation can be made. This compensation money would come from federal funds appropriated by Congress.

Since everything we're discussing involves declared emergencies, officials must move cautiously in deciding who will be vaccinated or treated and when. The frequency of complications and death from a vaccine versus the probability of contracting the infection without vaccination must be considered—which doesn't appear to be the case in the military's anthrax and smallpox vaccination programs.

On the other hand, the ring vaccination to eliminate smallpox and the antibiotic strategy following delivery of the anthrax letters appear right, because the probability of contracting the diseases was very high and outweighed potential complications and vaccination and drug cost. Population size is also a consideration, because the larger the population the greater are the risks for serious adverse effects.

Conversely, if a disease threat is real but small, not everyone should be vaccinated in advance. The necessary corollary is that the Strategic National Stockpile must be large enough and well enough coordinated with emergency response networks that vaccines and other countermeasures can be delivered rapidly when a small threat suddenly looms large.

Let's consider what different strategies might be applied in foreseeing the hostile use of smallpox. Experts argue three different positions, but none is without drawbacks as we've noted:

VACCINATE EVERYONE NOW.

- Based on data from 1968, a year when more than 14 million Americans were vaccinated,[41] we could expect 12,000 complications and 190 deaths, a higher rate now because of the prevalence of people with weakened immune systems who have had organ transplants or who have AIDS.

VACCINATE EVERYONE ONLY AFTER A FIRST CASE HAS BEENREPORTED.

- Inoculating 300 million people in as short a period of time as a month would be nearly impossible, even under the near ideal conditions that rarely exist.

RING VACCINATE AROUND REPORTED CASES, THE SPECTACULARLYSUCCESSFUL STRATEGY OF THE 1960S.

- We now travel frequently and with ease—even, it might be argued, unstoppably—and that would make it nearly impossible to quarantine infected people within protective rings. Consider the case of the young American attorney suffering from a dangerous form of tuberculosis who managed to elude health authorities for days while traveling on two continents,[42] then imagine trying to stop every one of hundreds or thousands of infected people trying to flee or to get home.

One Bug, One Drug

The crafters of BioShield 2004 inserted one more item whose four words just about guarantee that most drugs being developed for the Strategic National Stockpile will never do most of us any good at all. Yet these same four words will keep most major drug makers out of the field for lack of interest. No one quite intended it that way.

The culprit is a proviso that countermeasures developed for the Stockpile have "no significant commercial market."[43] The Strategic National Stockpile was envisioned under BioShield 2004 as an array of drugs and other countermeasures that, like a shield, would be ready and waiting for deployment against a bioweapons attack. While we have some concerns over Stockpile strategy and management, it does represent one of the important accomplishments of the U.S. biodefense program, potentially protecting us from natural pandemics and the more probable localized, small bioweapons attacks.

However, when considering potential drugs for inclusion, the Secretary of Health and Human Services was instructed to *consider* excluding those that had that significant commercial market. Unfortunately, "consider" has been taken as a mandate for exclusion. Presumably the intention was to prevent wealthy drug companies from profiting on countermeasure development by selling their publicly funded protectives for other uses at a large profit. But that also means big drug companies [44] would not be interested. First, their likely sales would be one-time-only to the Stockpile, bringing in at most a few hundred million dollars. By contrast, their hoped-for *annual* sales for a single drug start at hundreds of millions and range into the billions of dollars for treatment of chronic conditions, such as high cholesterol and depression.

The "no significant commercial market" provision has led to a countermeasure development strategy known as "one bug, one drug" — a strategy supported by the National Institute of Allergy and Infectious Diseases for research on bioweapons agents as well. "One bug, one drug" means that any given countermeasure should address only one single bioweapons agent. This prevents drug companies and academic researchers alike from using BioShield 2004 and earmarked NIAID dollars to develop just the kinds of broad-spectrum treatments that would work not only against biowarfare agents but also against such major public health threats as antibiotic-resistant staph infections and TB. Those drugs clearly would have a large commercial market—so they're out. Now anyone interested in developing a drug against anthrax or plague *must be nearly certain it works against nothing else* if BioShield 2004 funding is to be used.

Finally, if you were developing a broad-spectrum countermeasure that would fight anthrax and a range of other bacteria, you would do it most quickly, safely, and cheaply using a model organism such as *E. coli.* For a drug that could work only against anthrax, on the other hand, you would need to work—slowly, carefully, and expensively—in the dangerous *Bacillus anthracis* itself. Plague? Stick to *Yersinia pestis.*

Help on the Way?

One major improvement in BioShield 2006, titled the Pandemic and All-Hazards Preparedness Act,[45] is its focus on smaller biotechnology companies rather than large pharmaceutical firms, because the so-called Big Pharmas need too much in the way of incentives to come to the table. The biotechs are interested in BioShield funding, because a one-time sale of a few hundred million dollars is big to them.[46] And they may be better poised to develop innovative countermeasures. Another improvement—and a big one: it encourages broad spectrum countermeasures with commercial potential.

There are other signs of hope. A few years ago, the Department of Defense took the lead by creating a program called Transformational Medical Technologies Initiative, which funds basic research into molecular mechanisms of infectious diseases.[47] The program's goal is to discover pathogen pathways that might be targeted by new, broad-spectrum antibiotics and antivirals. This is precisely the approach the entire federal government should be taking, and finally now may take.

NIH has just published well-considered and coherent short-, medium-, and long-term strategies for biodefense countermeasure development,[48] in which basic research on new antibiotic strategies and broad-spectrum countermeasures are weighed appropriately. While the focus is biodefense, because that's where congressional funding for the program might be secured easily, such countermeasures will have important use in the other diseases, which are the bigger public health threats.

Essentially this strategy aims for "an integrated, systematic approach to the development and purchase of the necessary vaccines, drugs, therapies, and diagnostic tools for public health emergencies," according to the NIH's parent Department of Health and Human Services.[49]

The Public Health Priority

We need to shift far more resources toward natural infectious diseases and other pressing public health issues. Most strategies and countermeasures so developed would find immediate use in biological weapons defense. Multidrug-resistant infections are expensive to treat, and they are becoming ever more deadly. For example, a victim of MDR *Staphylococcus aureus* costs about $48,000 a year to treat. MDR *S. aureus* accounts for 50 percent of hospital-acquired infections—three hundred thousand cases per year and growing. This single, multiple antibiotic – resistant infection costs about $14 billion every year to treat, and it causes many deaths as well.[50]

The converse of the broad-based approach doesn't get us very far. Narrowly focused, "one bug, one drug" research does not speak to our most pressing needs and may never protect us at all. For example, monoclonal antibody countermeasures—the main focus of the "one, bug one drug" approach—circumvent drug resistance and kill bacteria, so they may provide an interim deterrent to a bioweapons assault. An enemy may decide not to launch an attack that will be neutralized. But monoclonal antibody drugs are very expensive, must be injected, and may have short shelf life in Stockpile storage. By contrast, so-called small-molecule antibiotics are generally inexpensive, stable and can be taken orally.

BioShield 2006 provides funding for pandemic and epidemic countermeasures as well as bioweapons and recognizes that protection against natural diseases must be part of biosecurity. However, it only includes public health emergencies, not traditional public health problems, so new legislation should address ever-present threats. Our biggest need is for a large increase in funding for annual and emerging infectious diseases. They are always with us. New funding should support academic and company infectious diseases research and development as well as intramural NIH research in programs that were in place long before 9/11. International public health threats should be attacked as well.

We need to take the high ground, the international approach. Biosecurity at home requires biosecurity everywhere. An epidemic of a highly contagious disease or a bioweapons attack in a remote corner of the world is an immediate threat to the United States. BioShield and related legislation should require strategies for international cooperation and collaboration.

The best way to develop effective countermeasures is in a climate of rapid, free exchange of discoveries and problems. Legislation also should include provisions to encourage worldwide distribution of countermeasures. A rogue nation planning to build a bioweapon would find it less attractive knowing countermeasures would be available to anyone. And our recognition that we are all in biosecurity together will improve world public health and goodwill.

BioShield 2006 opens the door to international cooperation by *allowing* funding of projects with highly qualified foreign nationals, as well as providing leadership in international programs and holding meetings that include international agencies.[51] Such cooperation is not required, but it should be—and was, in an earlier version.

That Senate bill could not find enough support for its mandated international collaborations.[52] It would have created an advisory committee of international public health, infectious diseases, and medical groups including the WHO and the International Committee of the Red Cross. It also called for "global distribution of [medical] countermeasures." Its language could hardly have been more accurate: "Experience with infectious diseases like HIV and malaria in developing countries illustrate[s] that [the] United States must work with developing countries to plan for the delivery of products developed under this act."[53]

Not that this Senate bill was perfect. It too often called for secrecy and provided for various competition-free, costly incentives to industry for countermeasure development. We must balance transparency and secrecy.

Coming up with examples that illustrate when either transparency or secrecy ought to win out is easy. Say you have discovered a mechanism of a bacterium's infection that makes a perfect target for antibiotics to attack: publish this quickly. Now imagine you have found a way to *cripple* an antibiotic that targets just that mechanism: keep it under wraps. Overall, however, openness should rule the day, and that means developing transparency through international cooperation. But transparency begins at home. Compliance with the Freedom of Information Act should have been mandated in BioShield legislation.[54]

As enacted, the law requires withholding all information that would be available under the Freedom of Information Act if it "reveals significant and not otherwise publicly known vulnerabilities of existing medical or public-health defenses against biological, chemical, nuclear, or radiological threats."[55] Hiding vulnerabilities to the chemical, biological, radiological, and nuclear threats may not always be a good idea, even if it seems so at first. How will you get the cadre of creative academic scientists to address vulnerabilities if they are unaware of them? Furthermore, "vulnerabilities" are ill-defined, leaving a large loophole for arbitrary and unreasonable withholding of information from the public.

The default position for whether to make information public or keep it secret should be the FOIA requirements. Not so under this law. Says Alan Pearson, "As written, the legislation could ... prevent the results of fundamental research from being released. This directly contradicts the longheld federal policy of maintaining openness in fundamental research."[56]

Pearson quotes from the government's own National Security Division Directive (NSDD 189) that states: "Our leadership position in science and technology is an essential element in our economic and physical security. The strength of American science requires a research environment conducive to creativity, an environment in which the free exchange of ideas is a vital complement."[57]

Pearson concludes, "We can't afford to be stuck with less than effective medicines. Yet, by promoting secret medicine, the current legislation threatens just that."

We need to push for much-needed stronger oversight. Projects in BSL-*3* and-*4* labs are dangerous by definition. That means they need mandatory *proactive* oversight—oversight that begins ahead of the projects. Furthermore, the decision to begin a dangerous project cannot be left to scientists alone. Scientists can be motivated by the same worldly ambitions as anyone, and, as we have seen, they, too, can miscalculate risk, develop a false sense of security, and just plain foul up.

Addressing real public health threats and improving international transparency and cooperation must be the focus of our efforts, marking an enlightened approach to biosecurity.

Not by Countermeasures Alone...

Jonathan Marks identifies three models for public health emergency preparedness: the single-hazard approach, all-hazard and multi-hazard approach, and, finally, improving health care infrastructure and access.[58] Regarding countermeasure development, the one bug, one drug approach to bioweapons agents and the single-minded emphasis on pandemic influenza represent the single-hazard approach. The strategies do almost nothing to strengthen everyday public health. The development of multi-spectrum antibiotics, which we urge, would be classified under multi-hazard, but that is only a small part of what is needed for true preparedness.

In Marks's classification, multi-hazard approaches include "systems to monitor and track the spread of biological agents and infectious disease, enhanced laboratory response capabilities, incentives for health care professionals to report to work in the event of an emergency, back-up facilities, and supplies of medical equipment to meet demands for surge capacity in the event of a public health emergency."[59] There has been some progress developing systems to monitor pathogens and identify them quickly in the laboratory. Federal legislation has been passed—albeit, the wrong kind in our opinion—to protect health care workers from lawsuits. However, we may have actually backslid in improving our ability to handle the surges in demand that would follow either a sizable bioweapons attack or sweeping flu epidemic.

We are woefully behind in developing this surge capacity, as pointed out in a recent Trust for America's Health Report.[60] A mere $320 million of the nearly $50 billion spent on biodefense from 2002 to 2008 was to aid states in biopreparedness. That works out to a piddling $900,000 per state per year. Compare this to the $3.3 billion spent by BioShield on single-hazard countermeasures.[61] Worse, whatever has been awarded to hospitals in biodefense funds more than likely has been taken back through federal funding cuts to states and cities.

This dangerous lack of surge capacity is underscored in frequent news reports of ambulances being turned away from overcrowded and overtaxed emergency rooms, often during ordinary seasonal flu outbreaks. On top of that, in a surge caused by a severe infectious disease outbreak, requiring new levels of protection would make even the most routine hospital activities arduous. In a recent presentation, the associate director of the Massachusetts Nurses Association warned that it would be extremely difficult to perform such simple tasks as inserting an IV when clothed in a hazmat suit—the very suits pictured on the title page of the Trust for America report. Mary Crotty told the Boston City Council that even getting to the hazmat suits could produce a bottleneck in one Boston-area hospital since they are locked in a closet, with only one key in the hands of the supervisor, who works the 9 to 5 shift.

As Marks suggests, without sweeping measures including universal access to health care "new vaccines and therapies will ... help just a few. If we are to have any hope of exerting control over a global pandemic—whatever its source—helping [only] the few is an option that none of us can truly afford."[62]

A Profession on the Roller Coaster

Public health was a highly respected profession until the 1920s, according to David Ozonoff, a longtime public health professor.[63] But then it came "completely under the thumb of clinical medicine" and public health became "a very low-status place to work." When Ozonoff began medical school in 1962, going into the field meant "you were going to read TB films in the South Bronx. The smart kids went into internal medicine and psychiatry." But then the social ferment of the'60s took hold, and suddenly "a lot of medical students ... wanted to combine their occupations with their political ideology. Public health was the perfect place for that to happen."

Unfortunately, the renaissance was short lived.

Ozonoff recalls that the field went into steady decline under the Carter, Reagan, George H.W. Bush, and Clinton administrations. Carter was for small government, Reagan's neoconservative advisers did not like publicservice government programs, and Clinton pushed welfare reform "a step backwards in providing the social safety net for the American public."

Then came the George W. Bush administration. "Under Bush Two, we had been on the edge of the precipice and he just gave the push that sent us over the edge."

How was all this reflected in infectious disease control? Starting in the 1960s, many experts began predicting that combating such ills would soon be a thing of the past. Even as antibiotic resistance began to appear, drug companies thought they could outrace the bacteria simply by modifying existing antibiotics, so they began to aim R&D away from antibiotics and toward more lucrative markets.

Then came the AIDS pandemic. Although HIV is a virus, its elusiveness demonstrated that our main infection defenses—in this case, vaccines—could not always protect us. Soon the emergence of ever-more resistant bacteria confirmed that our defenses against virtually all infectious diseases are vulnerable to the evolving strategies of microbes.

At first, however, the events of 9/11 and Amerithrax seemed like a way to restore funding and status to public health. Ozonoff notes that public health officials saw the biodefense push as a way to restore lost funding. But when the government used bioterrorism and biodefense as the leverage for public support, public health was "neutered to a large extent," and its new funding prospects evaporated. He says officials "thought they were going to use the biodefense money, but in fact the biodefense people were using public health."

Public health itself now is in a state of disaster, says Ozonoff, because now instead of being "under the thumb of the medical profession" it is "under the thumb of public safety," whose military-like organization often demands secrecy.

The field's reorganization on the public safety model means vertical chains of command instead of the horizontal organization under which it had thrived, in which maternal and child health, substance abuse, and vital records were all on a par. Now a major public health mission is to protect the population against terror attacks. "And in order to do that, you have to buy into the entire military organization that's being used to organize against terrorist attacks—the incident command system."

How does that affect the rest of us? First, if there is an infectious disease outbreak, the head of the health department might not even be the "incident commander." But the real changes go far deeper. As Ozonoff puts it, "When public health works, nothing happens. In the incident command system, nothing works until something happens." Thus, "instead of being preventive, public health has become reactive."

And then there's the secrecy. "Once you agree that there are things that are too important or dangerous to tell people about, all the work that you've done is no longer public health. Public health has got the word 'public' in it. It means you take this stuff ... to the world scientific community. It doesn't mean you keep it secret."

These problems permeated even the CDC, the "jewel in the crown of public health," where many professionals, including scientists, "have been trampling each other at the exits to get out."

Ozonoff lamented, "They've lost their institutional memory," and, "they've lost sight of the mission." The problems started at the top, he said, where Julie Gerberding, then director of the CDC, pushed "the catastrophic reorganization"—that is, the reorganization that has made fighting terrorism a major CDC mission.[64] It remains to be seen how the CDC will operate under the new administration.

The public health roller coaster has been plummeting. Have we finally bottomed out?

CHAPTER NINE

Down to Grass Roots

Wake-Up Call

Federal officials have frightened us over bioterrorist-crafted weapons much less likely to cause massive sickness and death than the next pandemic—and that pandemic may be one of our own devising in a bustling urban neighborhood like Roxbury in a sophisticated, high-tech city like Boston, where biotechnology contributes much to the economy and to our future.

Nothing so illuminates the lethal dangers that may soon be hidden from public scrutiny as the government and Boston University's decision to build a BSL-*4* and BSL-*3* laboratory complex [1] for investigating the most dangerous pathogens as part of BU's medical center, which is surrounded by three big neighborhoods that are home to a hundred thousand Bostonians. Here is where overarching federal policy runs up against grassroots concerns over health and safety. Construction is being pushed forward following utterly incomplete risk analysis.

It would be tempting to write off the BU case as a singularity with little relevance outside its metropolitan area. Nothing could be further from the truth. In terms of what is planned, the causes of the protest, the arguments being used as justification, and the project's relevance to national and international concerns, Boston may serve as the strongest argument for oversight and transparency everywhere.

Many communities are probably unaware of what is going up down the street or what is already being carried out in a nearby BSL-*3* lab. Because labs working on select agents, which include bioweapons agents, must register under the Select Agent Rules at either the CDC or USDA, in theory the government should know where the labs are and what they are doing. But it is unclear whether any government agency has compiled or even looked at the data. Even aggregate data, if it has been compiled at all, has not been released to the public.

Therefore, and perhaps most important, some Massachusetts public officials alerted to Boston University's activities are trying to legislate ways to keep all labs in the state under public scrutiny. In the absence of national regulation, state and local legislation can serve as a seed for national and even international regulation. In fact, legislation similar to that in Massachusetts is being proposed elsewhere.[2] Transparency, proactive oversight, and no-notice inspections are the keystones of these pro posals.

Protests against construction of the top-containment lab began to shape up, as most do, with even an informed lay observer likely uncertain which side was right. Residents marched with picket signs in a nondescript parking lot near the Boston Flower Market that was destined to house such microscopic flora as the hemorrhagic viruses Ebola and Marburg. On the other side, Boston University assured them that no living thing could escape what would become the paradigm of brilliant strategies for risk containment. The labs were endorsed by Boston's mayor, Massachusetts's two U.S. senators, and the district's congressional representative, all of whom were strangely silent on the subject of risk.

Anyone might think this was a case in which inexpert neighbors simply did not grasp the true safety level that would be built in, or the matchless work the lab would carry out to the benefit of mankind. But it was not so this time. Powerful forces soon lined up on their side.

Active support came from 165 leading scientists, including two Nobel laureates, who all signed a petition against the project being pushed by some of their own colleagues.[3] They argued that a "Level 4 laboratory should not be located in this densely populated urban area." That was a couple of years ago, and their stand has not changed. Legal muscle came from a number of nonprofits and law firms.[4]

At least one scientist initially supported the project but came to oppose it. David Ozonoff of BU's School of Public Health believed the plan was to build a laboratory that would study emerging infectious diseases. Then, at a meeting, community members and some of his own colleagues showed him what was really afoot: "It was a biodefense research and development laboratory," he says, with just a few crumbs for public health. "I did a 180." Although he does not oppose the location, per se, he also was angered by the "high-handed, arrogant way" in which the planners ignored community concerns by locating yet another undesirable use in a neighborhood that already had too many.[5]

As noted, the number of operational BSL-*4* labs in the United States is approaching fifteen.[6] The case can be made that one BSL-*3*/BSL-*4* agricultural complex should be built on Plum Island off Long Island, New York, to replace the crumbling and obsolete BSL-*3* labs there now. And that probably should end such projects.

The reasons no new BSL-*4* labs are needed are simple. Menaces like Ebola, Lassa, and Marburg viruses—all of which require BSL-*4* security—are not public health threats anywhere in the world *except* the African countries where they are endemic. Even in Africa they are minor threats compared to AIDS, malaria, tuberculosis, and a host of other diseases. So there is no need for new labs to study them when well over forty adequate facilities already are in operation here and abroad.

In an effort to downplay its biodefense goals of the lab project, Boston University recently has emphasized an emerging infectious disease focus for the labs, but the most pressing of these illnesses require only BSL-*3* protections, if not simply BSL-*2.* Emerging exotic diseases like dengue need attention.[7] Conditions in many developing countries where large portions of the population have weakened immunity from AIDS and cities are rife with overcrowding and poverty, plus the world population's global mobility, are increasing the incidence of many once-rare diseases. That said, our thousand-plus BSL-*3* labs are more than enough to carry out such inarguably important investigations.

Genuine public health threats, including HIV/AIDS and MRSA, require only BSL-*2* containment—the level found in countless university infectious diseases labs. Only particularly dangerous experiments with the HIV virus require BSL-*3* containment. Why such low-level containment for such fearsome microbes? Because the diseases they cause already are endemic in the United States, or they are not particularly infectious in a lab setting, or they are not particularly contagious even if infectious. By contrast, studying the 1918 influenza virus does not require complete BSL-*4* security—although with its record of some 40 million dead it certainly should, in our opinion. Research on highly contagious viruses such as SARS and the 1918 pandemic flu should be carried out only at the CDC, if at all. Placing these agents in labs throughout the country as we are now doing is extremely dangerous, as it increases the chance for escape and an epidemic.

Even anthrax is not normally fatal to humans, and there are both antibiotics and a vaccine against it, so studying this bacterium usually requires only the protections of BSL-*2*—*except for aerosol experiments on weaponized deadly strains, work with large quantities, and experiments that involve infecting large animals,* all of which are dangerous experiments.

In that emphasis lies a major reason why so many biosecurity experts are alarmed by our ongoing highest-security lab development. Boston's and other BSL-*4* labs might well be researching bio *defenses* against weaponized anthrax, Ebola and Marburg, but it certainly will not look that way to real or potential enemies. Such work looks like a cover for offensive bioweapons development, and because of the secrecy increasingly shrouding bioweapons research, there will be no way for us to prove otherwise.

What began as a neighborhood-government confrontation in Boston has emerged as a microcosm of biodefense in America in the new century, as the university and the federal government muscle ahead with highsounding purpose that combines biodefense with emerging diseases investigation.

Boston University ensures political support by claiming the project's necessity to area bioscience and it answers opponents with platitudes that ignore the central issues we highlighted above.

Roxbury is a minority community that has been subject to a number of undesirable and environmentally suspect projects, so the fight for "environmental justice" has been long and continuing. Led by Klare Allen, who says her life as an organizer began when her family was left homeless after an illegal eviction,[8] the efforts against the BU lab expanded to include law firms, medical organizations, local peace groups, scientists, and biosecurity experts.

Boston and the surrounding communities are sophisticated cities, the homes not only to Boston University but also to Harvard and its professional schools, MIT, Tufts, Boston College, Northeastern, and many more. No U.S. metropolitan area has a better record of strong support for the life sciences. So the drama playing out there does not represent Luddites striking out against change.

Finally, though brilliantly designed, as any high biosecurity – level laboratory might be, no security system is foolproof. A disgruntled employee could deliberately blow up a lab from inside with a powerful explosive like nitroglycerine prepared from glycerine and nitric acid, two common laboratory chemicals. And the roster of dangerous, unintended accidents is growing right along with the increasingly dangerous activities being conducted.

The representatives and scientists pushing for development say the $128 million labs will bring both revenue and prestige to the city. They note that NIH's risk assessment showed the operation would be safe and Boston was an even safer location than one in the suburbs or rural areas, and claim that the project is vital to Boston's economy and its bioscience communities—major pharmaceutical companies, biotechnology companies, and academic and other noncommercial labs.

Challenging the safety claims, the 165 opposing scientists note that the BSL-*4* research "will involve work with some of the most dangerous biological organisms, viruses and toxins that can cause deadly diseases for which there are no known cures," and that "there can be no guarantees that there will be no accidents that might lead to the release of deadly, air-borne pathogens through the failure of safety systems or when they are transported through city streets." And if bioterrorism is really the fear driving the project, they added, "the laboratory might also become a target for intentional acts of violence."[9]

How, then, did the NIH give such a clean bill of occupational health to the labs? In part, at least, by failing to consider all the risks such work presents to both lab employees and the public. Downplaying real threats while overstating imagined ones has been the tragic hallmark of U.S. government policy and the intelligence used to justify it since 2001. In this case, of all the ways dangerous infectious disease agents might escape from a high-biocontainment laboratory and the many kinds of agents that could escape, the NIH assessment for the BU labs modeled only one scenario—one in which population density of the community surrounding the lab would not matter.

Next, let's look at the ways a public health disaster might emerge from a high-biocontainment lab, via events we have already looked at:

- An aerosol is accidentally released into the neighborhood.

- A lab worker who has become infected heads home after work and infects strangers, who cannot be easily traced, on a crowded subway.
- The microbe is accidentally carried out on a worker's clothing or on vermin.
- Solid or liquid waste handling systems leak or their sterilization fails.

That's at least four different pathways. And what sort of microbe might be released in such an accident? As we know, it might be:

- Noncontagious, like anthrax. It can kill large numbers of people who breathe it in, even miles from the release, but it cannot be transmitted from person to person.
- Mildly contagious, like Ebola—hard to believe if you've read the shock stories, but true. Though as deadly as portrayed, Ebola is minimally contagious. Contact with infected bodily fluids usually is required for transmission, so those living or working close to victims are generally the only people at risk.
- Highly contagious. SARS and the 1918 pandemic flu virus are both highly contagious and quite deadly. In fact, though far fewer people have died from SARS, this viral infection killed a far higher percentage of those infected than did the 1918 flu.

That's three different levels of contagion.

For a complete risk analysis, it's obvious that each type of agent must be considered against each type of escape. As we've categorized them, that makes four times three—or at least twelve—different scenarios. The NIH chose one: a single researcher is infected at work with a mildly contagious but quite deadly virus and brings the infection home. Notice that this is the threat in which only family members, health care providers, and others in intimate contact would be at significant risk. And notice that it is a scenario in which population density does not matter, since other people would not be exposed. The NIH asked only the question that would give the answer it wanted—a dangerous way to approach a serious public health issue.

The analysis did not consider an accidental aerosol release in an adjacent neighborhood like Roxbury, South Boston, or Boston's South End, where the population density is ten to a hundred times that of a suburban or rural site. In this scenario, population density might well be a critical factor. A Boston lab worker infected with a highly contagious disease who transmits it by coughing or touching public doors and railings would present a much greater risk of infecting many strangers on a subway or street than if he or she worked in a suburban or rural lab; not only would he or she contact fewer people directly, but it is more likely that the employee would drive to work.

Indeed, a National Research Council committee, convened at the request of the Commonwealth of Massachusetts, has found the NIH draft risk analysis for the proposed lab to be "not sound and credible."[10]

The BU medical center BSL-*4* lab may still be approved, but the NRC finding is a setback. Armed with the NRC report provided by the Massachusetts Office of Energy and Environmental Affairs, "the Supreme Judicial Court agreed that the state's environmental approval of the South End lab, granted by the Romney administration, was "arbitrary and capricious."[11] According to Douglas Wilkins, one of the key lawyers for the lab opposition, for BU to move forward with a BSL-*4* lab requires state permits that it cannot obtain. "I think they're dead in the water at the moment," he said.[12] But BU is not giving up, and the lab still could eventually be approved.

The lab controversy is now back in the hands of the NIH, which in March 2008 appointed a Blue Ribbon Panel of "experts in infectious diseases, public health and epidemiology, risk assessment, environmental justice, risk communications, biodefense, biosafety, and infectious disease modeling."

The panel is charged with addressing public safety concerns and will "advise the NIH in determining the scope of any additional environmental risk assessment that the agency will conduct."[13] It may take the panel two years to complete its work.

If the lab were approved, would Boston garner prestige? It's hard to believe that a city that is home to so many of the nation's leading universities and research centers would garner laurels from one more lab. The biotechnology industry was spawned and has flourished in the Boston and San Francisco metropolitan areas without a single BSL-*4* lab. And importantly, nearly all of Boston's surrounding municipalities have passed resolutions opposing the project, from Cambridge, home of Harvard University and MIT, to Brookline, Arlington, Somerville, and Newton.

In terms of revenue, Boston University is nonprofit, so its buildings and operations pay no taxes. What about income from the biotech and pharmaceutical industries? Here is the most likely forecast: of the ten or so major pharmaceutical company facilities, the more-than two hundred biotechnology companies and dozens of noncommercial infectious disease research labs in Massachusetts, only a handful will ever use the BSL-*4* lab at Boston University.

We've already discussed how hard it has been for the federal government to lure Big Pharma into countermeasures development. Even beyond that, there is a lot about BSL-*4* work for a pharmaceutical company to dislike, such as arousing suspicions that it is developing bioweapons, and liability if it develops non–FDA-approved drugs for emergencies.

But one negative trumps all others. For reasons ingrained in their nature, pharmaceutical firms abhor regulations and inspections, believing they already are overburdened with both. Working in labs with select agents would pile on new layers of regulations and inspections mandated under Select Agent Rules, and, worst of all, any mishap might bring on the FBI.

All that for efforts that would reward a company with a fraction of the income of a prescription sleep aid.

Acting Locally

In response to the many local concerns, the Boston Public Health Commission adopted the Biological Laboratory Regulations, in order to "strengthen safety oversight of biological research labs across the city."[14] These are the rules the BU labs would have to live by, and they go part of the way toward overcoming neighborhood fears. They could be the bellwether for how all Americans live with these neighbors, for better or for worse.

Under its new rules, the commission may conduct no-notice inspections of high-biocontainment laboratories, a key provision. Managers of a lab that could be visited at any time with no notice must weigh the risk of getting caught without time to cover up an unsafe or noncompliant activity—and as we've seen, there have been plenty of precedent examples of such activity. Further, the Boston regulations forbid any research that might turn a high-risk agent into "a principal component of a biological weapon," or that would even significantly aid in its construction.

But the Boston regulations have potentially serious flaws. While the institutional biosafety committee will include two community members, they will be chosen by the institution, so whether they will have the community's interests or the institution's interests at heart remains to be seen. Many of the regulations' provisions are not clearly written. Most important, it is not clear if the commission can stop unsafe experiments before they are begun—that is, whether it will have proactive oversight.

One final provision almost certainly will put the health commission on a collision course with the federal government. The Boston regulations forbid secret research on high-risk agents without full disclosure to the commission. The high-biocontainment labs may well be forced to carry out such classified defense research, despite Boston University's public assurances that they won't, because complying with such requests is part of the funding agreement with the NIH. The lab personnel would have appropriate security clearances, but the commissioners and public members probably would not. Something will have to give.

However, the Boston Public Health Commission is not alone in its efforts to force transparency in high-containment biological research. State representative Gloria Fox is pushing a bill that would mandate many of the same rules for openness and disclosure as the city commission's, and naturally would do so statewide. And the Fox legislation overcomes pitfalls in the city provisions.

Her bill calls for no-notice inspections and would regulate and oversee all research in BSL-*3* and BSL-*4* labs. A major provision would create a state Biological Agents Registry. There, all BSL-*3* and BSL-*4* labs *and* all the pathogens they hold in Massachusetts would have to be registered. Now the state has no idea where the labs are or what pathogens live among its laboratory denizens—an ignorance shared by all fifty states and the federal government.

Finally, the two community members on the institutional biosafety committee are appointed by the state and local boards of health, so they should represent the community's interests and the bidding of the boards of health. The bill requires the IBC to review and approve all projects in high containment labs before they get underway—the critical proactive mandate.

When will the Fox bill become law? It has been bottled up in committee, brought back and revised, bottled up again this and last year, and it didn't come to a vote of the full legislature in 2007 or 2008. Hopefully it will not suffer the same fate as the Protocol to the Biological Weapons Convention, for its proposed mandates reflect some of the best provisions of that international effort.

For more than five years, arms control experts and diplomats from over fifty nations wrestled with a transparency and oversight protocol to the BWC. The protocol rested on three pillars: declarations of dual-use facilities by the nation housing them, short-notice transparency visits, and challenge investigations that would be more intensive than the simpler visits.

Later we'll turn to the fate of the important Biological Weapons Convention protocol and consider why many of its provisions must be revived, some in local legislation. The Massachusetts legislation contains many provisions of the BWC protocol: short-notice inspections parallel the BWC's transparency visits and challenge investigations; a Biological Agents Registry similar to the declaration of dual-use activities; and finally, institutional biosafety committee meetings open to the public and local representation on the IBC, also akin to the BWC's transparency visits.

Loading the Magic Bullet

Projects in BSL-*3* or BSL-*4* labs are dangerous by definition. If they were not, these labs would not be called for. The BU case shows why the public's right to know needs to be protected and enhanced. But because of the dual-use nature of so much biological research and development, the painful choices facing scientists often involve far more complex issues than the risks of spills or escapes. Very often their choices involve what psychologists call "conflicted intentions." Let's consider a realistic scenario that could easily play out anywhere in America—or the developed world—now or in the near future. And you are in the driver's seat.

A biotechnology company is on the verge of what might be a double bonanza—a promising therapy for an extremely deadly cancer whose success could translate into explosive corporate financial growth. Magic Bullet, Inc., has isolated two proprietary compounds from different microorganisms. Each compound alone is completely safe when injected into mice at typical drug dosages. But when the two are injected together, they form an extremely toxic combination—even more toxic than botulin—that kills the mice.

Magic Bullet would like to harness this harmless/deadly feature to treat colon cancer. Its scientists hope to target the two compounds to the surface of colon cancer cells, and only those cells, by tagging each with a peptide that binds to a protein found only on the surface of those cells. Such unique surface "signatures" are common in cancer and in many types of healthy cells.

If this proves successful, a patient might be injected with the two tagged compounds one at a time so the first compound clears from the blood and is present only on the surface of the cancer cells before the second compound is injected. However, when the second compound reaches the target on the cancerous cells, the two combine at toxic concentrations. Death to the cancer, life to a patient's normal tissues. The crucial missing pieces are finding the unique surface molecule on the colon cancer cells and the peptides that carry the compounds to the unique cell sites.

That's where you come in.

You have just received a PhD in molecular cell biology, and your thesis involves just the skills Magic Bullet needs to test its idea, first in cell culture, then mice, and finally in humans via the three stages of clinical trials. You are quite excited about the prospects of these compounds and Magic Bullet's strategy for treating cancer, so you accept a good

R&D position in the company. What an opportunity! You could ultimately accomplish great good by doing the science you love, and you might even become wealthy doing it. Of course, none of this will come quickly, as you well know after six years of effort earning your doctorate.

Now after four years with Magic Bullet, you strike pay dirt. You devise a good peptide that targets and binds to that specific protein found only on colon cancer cells. Each of the two safe compounds can be bound in separate test tubes to the peptide. Next you perform the needed experiments on cultures of cancer cells and those of normal cells, demonstrating that only cancer cells are killed. The candidates for the peptide/toxin drug are dubbed "MagiToxA" and "MagiToxB." You work rapidly now, demonstrating that you can completely cure mice of colon cancer. Everyone at Magic Bullet is ecstatic and would like to begin clinical trails quickly.

But a storm cloud rises on the horizon. Biochemically isolating and purifying the two compounds from their finicky host microorganisms is both expensive and time consuming. It might cost $100,000 per patient to manufacture MagiToxA and MagiToxB—awfully expensive, even for a cancer drug. For the phase I clinical trials that will test safety on only one hundred carefully selected patients, that means the company must spend $10 million just to produce sufficient drugs. The firm simply doesn't have that kind of cash, and you have not achieved enough to draw new investment, because venture capitalists in the current bad investment climate are only funding drugs that have, in the current parlance, "touched" humans.

Magic Bullet is burning $5 million a year on R&D; now it has just $5 million left in the bank. That means the firm must delay clinical trials until it has an inexpensive means of obtaining the drugs through genetic engineering into easy to grow *E. coli,* and that is not likely to come quickly. Company managers believe that successful phase I clinical trials would enable them to raise the necessary financing to continue operations, but they face a common catch-22. They cannot get to the trials without financing, and they cannot get financing without showing success in safety trials. If their goal is blocked, so is hope for the cancer victims who might benefit.

They turn to you. Praise from the scientific community over your dramatic experimental results might yet attract investors, so management has asked you to publish details of the compounds and your experiments in a leading journal. Patents have been filed on MagiToxA and MagiToxB and the method of combining them, so there is little worry about competitors profiting from your discovery.

To raise much-needed cash now, management wants to offer the information in its proprietary toxicology-toxic substance database that it has developed over the years—from which the two compounds were first identified—through subscriptions, so others can use them to help guide their drug development and to identify toxic substances. Magic Bullet has also developed powerful bioinformatics programs to mine the databases, which subscribers will get in their package.

Then comes potentially the biggest boost. The military learns of the exceptional combined toxicity of MagiToxA and MagiToxB. The army offers Magic Bullet significant funding for a project to understand what makes the combination so toxic, then to make a single compound with that much toxicity. The stated motivation is to develop antidotes to this new class of compounds in case they are used against the nation in a chemical attack. Of course, the toxic compounds you help develop with this military funding might lead to better cancer agents in addition to allowing your successful colon cancer project to move into clinical trials.

What ethical concerns do you have at this point? The history of the twentieth century stands as prelude in considering the bright and dark possibilities. Most obvious are the dual uses to which the work of you and your Magic Bullet colleagues can be turned. Any of your toxins designed for cancer therapy could be used for a new class of very potent chemical weapons. And then there's the proprietary toxicological database amassed by the company and the software for mining it. Chemical and biological weapon developers should be expected among the first subscribers to the database, and they will appreciate the details published in the journal paper. And what about engineering the compounds into easy-to-grow *E. coli* strains? Is this too dangerous to do? What if the strains were stolen?

None of this alters the fact that this is very exciting research that holds great promise, but it is short of financing. What do you think you and your Magic Bullet colleagues will do? More to the point, what *should* you do?

Knowledge with Conscience

Unlike the Roxbury conflict, the Magic Bullet dilemma cannot be resolved by anyone but scientists themselves, and few of those entering life science labs have had any training preparing them for the prospects of hostile exploitation and dangerous experimentation in the new biology. Maybe more seriously, their seniors at the top of the decision tree, who would be most knowledgeable about consequences of a planned enterprise, usually also have had little or no ethical training beyond such immediate concerns as patient or research subject privacy and safety.

In a world of intensive international, national, and regional travel where nearly everyone is on the move, there are no "isolated mountaintops." We are all neighbors of biological laboratories and the beneficiaries—and potential victims—of dual-use developments. The kind of quandaries described here can affect the globe soon after they affect a neighborhood. Fortunately, many prestigious national and international organizations and individual scientists and policy experts have seen how urgently they must deal with the gap between top-level scientific skills and ethical reasoning when it comes to dangerous research.

A statement from the International Academy of Science, representing academies of the sciences from dozens of countries, offers a forceful summary of scientists' ethical responsibilities, opening with this from Rabelais, a writer usually associated with the ribald and satirical: "Knowledge without conscience is simply the ruin of the soul."[15]

"Scientists have an obligation to do no harm," the academy declares. "They should always take into consideration the reasonably foreseeable consequences of their own activities," bearing in mind "the potential consequences—possibly harmful—of their research and recognize that individual good conscience does not justify ignoring the possible misuse of their scientific endeavour." They must "refuse to undertake research that has only harmful consequences for humankind."[16]

Scientists and their organizations have not always been so forceful in weighing the outcomes of their research beyond its power to advance knowledge. Jeanne Guillemin noted in a recent article that her own preliminary inquiries indicate that "very few Western microbiologists have paid attention to the potential for harm in their work," adding that "their common characteristic is a reliance on scientific methods with no necessary moral component."[17]

In the twentieth century, tens of thousands of microbiologists worked in secret state programs that ignored international rules requiring them to protect civilians, Guillemin writes. "Very few of these BW scientists ever recanted their dedication to helping infect masses of civilians with anthrax, tularemia, plague, smallpox, and other diseases."

Many of the scientific and medical organizations that recognize the need for formalized ethical rules want a code of conduct developed, possibly embodied in an oath similar to the traditional form of the Hippocratic Oath that binds physicians to "first, do no harm."

Gigi Kwik Gronvall, an associate with the University of Pittsburgh Center for Biosecurity, points out that such codes are needed because responsibility for dangerous activities increasingly falls on individual scientists. "The ability to use biology for harm is no longer the province of teams of scientists and large budgets, but a possibility for a trained scientist working alone at the bench," she wrote in *Nature Biotechnology.*[18]

And Ronald Atlas, codirector of the Center for Deterrence of Biowarfare and Bioterrorism at the University of Louisville in Kentucky, said codes of conduct are needed "to prevent the life sciences from becoming the death sciences through bioterrorism or biowarfare." He added an important requirement that speaks to the need for ethical training among scientists: "What we really need to do is create a culture of responsibility."[19] Atlas's view of such a code would mean complying with the Biological Weapons Convention and avoiding any research that is clearly intended or is highly likely to facilitate use of biological weapons. Codes of conduct and oaths are appealing because they can easily be implemented institution by institution, so they represent small investments even if their effectiveness is debatable. Other approaches require more effort, cost, and cooperation.

Ethical education of scientists is sorely needed as well, so that they recognize ethical issues and understand their responsibility to respond. Ethically aware scientists may be our most important defense against hostile exploitation and ill advised dangerous experiments in the new biology.

And in group discussions at many universities, sociologist and philosopher Brian Rappert discovered that many scientists are unaware of important biosecurity issues.[20]

It is interesting that scientists, who are better equipped than the general public to gauge the seriousness of terrorists' use of bioweapons, sometimes conclude that the whole topic simply marks a political scare tactic. So ironically, it appears that the fear promoted by government officials may have backfired to cause scientists to underrate the threat posed by biological weapons even as the public overrates the threat of a large-scale bioterror attack.

Unlike nuclear physicists, who have seen and discussed the potential for great harm from their discoveries almost from the start, life scientists appear not to be anywhere near such an understanding. Requiring ethics education, establishing codes of conduct, and developing clear regulations of dangerous experiments and similar measures are essential steps to changing this dangerous situation. In concert they would constantly remind life scientists of their ethical and moral responsibilities.

Others biosecurity experts like Guillemin believe that only laws will prevent reckless behavior by some scientists. "The best hope for protection against biological weapons lies in the range of legal restraints that have been gradually building over the last several decades," she says.[21]

Blowing the Whistle

But whether through codes and oaths, ethical education or formal legislation, what would scientists do when they became aware of violations by others—whether at the behest of their companies or their countries?

The International Academy of Science statement takes a clear stand, and it is perhaps most significant for all of us, whether in or out of science: "Scientists who become aware of activities that violate the Biological [Weapons] Convention or international customary law should raise their concerns with people, authorities and agencies."[22]

They should, in other words, become "ethical dissenters," a term that translates more commonly as "whistle-blowers." However vivid the image, and however popular the figure in the news and entertainment media, the whistle-blower travels a difficult road. Yet we believe that such dissent by individual scientists will play a key role in our biosecurity—perhaps *the* key role.

Why do government and corporate scientists choose to remain silent about illegal or dangerous activities, even though their conscience tells them otherwise? Decades ago, a survey of 87 American whistle-blowers from both government and industry revealed that all but one experienced retaliation, and the more senior the whistle-blower, the greater the retaliation. That ranged from economic punishment to personal abuse.[23]

Whistle-blowers often lose their jobs and are sometimes even treated as though they are crazy. Besides fearing reprisal, these scientists often are under legally enforceable secrecy or confidentiality agreements, standard conditions for their employment, and violation of those agreements carries civil and sometimes criminal penalties. Financial pressure in companies and patriotism in defense labs also put pressure on scientists not to divulge suspect activities.

It is especially important that whistle-blowing scientists have special legal protection to protect them from retribution, slander, financial hardship, and lawsuits. They are in a good position to be aware of illegal or dangerous activities because of the important positions they hold in most companies and public institutions. However, according to Henri-Philippe Sambuc, an advocate of such strong whistle-blower protections, 95 percent of scientists and engineers work in enterprises such as industry and defense [24] where they have little to no independence—even though individually they have the power to do great harm at their fingertips.

Scientists now stand alone. They are not protected by unions or international law, as workers often are. In place of traditional unions it might become the role of major scientific organizations to provide legal and other support to whistle-blowers, even though the organizations themselves have no legal standing.

Scientists are working on means to protect them from identification and retaliation for taking ethical stands against government or other employers, using tools they developed for other communications as well as the latest devices of the Internet age.

One of the most intriguing models itself on Wikipedia, the online free encyclopedia to which anyone can contribute anonymously. Its name: Wikileaks. Its goal: to create "an uncensorable system for safe mass document leaking and public analysis. Our primary interests are in Asia, the former Soviet bloc, Latin America, Sub-Saharan Africa, and the Middle East," the Web site says, adding in what could be a critical addendum for this issue, "but we expect to be of assistance to peoples of all countries who wish to reveal unethical behavior in their governments and corporations."[25] The link comes from the Federation of American Scientists, which posts its own Web-based "Secrecy News" to broadcast information on government secrecy around the world.[26]

But many in the biosecurity and scientific communities have grave, well-founded concerns over Wikileaks. The site could be sued for slander for publishing wrong or hateful information that a court finds harmful. Confidential or classified information revealed there could provoke legal retaliation from companies or governments. Both possibilities mean that site submissions would need to be refereed by experts who could weed out wrong and harmful information and accusations. This would be a difficult and expensive undertaking, but a necessary one for the site to survive. Finally, site operators would have to take pains to minimize the risk that information shared over their network would be useful to bioweaponeers.

Recent efforts by the Center for Arms Control and Non-Proliferation might offer a precedent. The center considered creating a grassroots e-mail network to serve as an early warning system to put pressure on researchers and to establish informal oversight of possibly dangerous experiments. That network also could have served as an outlet for whistleblowers. The network or listserv would have operated along the lines of ProMED, the highly successful international forum used for more than a decade by disease specialists to immediately report information about suspected outbreaks, at a speed that can overcome the often-critical delays before official pronouncements are aired. Submissions to ProMED are refereed by experts before they appear on the site.

The center dropped its project, however—not because of legal concerns or refereeing problems but due to the unfortunate lack of interest in the bioscience community with dangerous and hostile exploitation of their own science. A refereed network can be costly for site maintenance and payments to referees, and given the lack of interest, the network was judged to be not cost effective.

In the end, there is no magic bullet that will bring us biosecurity. Every reasonable approach to increasing our protections must be explored, from rapid information exchange to international transparency to ethical education of scientists and involvement of the public. In Boston, we saw one end of the spectrum, local and individual ways of increasing biosecurity in the age of the new biology. But we must succeed as well in the national and international arenas, where it is essential to focus our attention and our political will.

CHAPTER TEN

Acting Globally

For all the foresight and good intentions behind the framing and passage of the Biological Weapons Convention in the 1970s, its goals have been eroded by fear and suspicion among its members, from the Soviet Union's acting on its belief that the United States was secretly violating the treaty to South Africa's attempting to develop deadly biological measures under cover to thwart its foreign and domestic enemies. Now fear and suspicion, the underminers of confidence and trust, have settled in here at home.

In 1986 and 1991, BWC members agreed to exchange information with one another to build confidence that everyone was complying with the treaty.[1] These confidence-building measures included information on defensive programs, high-biocontainment laboratories, vaccine manufacturing plants, and infectious disease outbreaks.

America has supported and generally complied with the measures, submitting the required annual declarations—up to a point. But some agencies skirted them by tinkering in secret with projects Jefferson, Bacchus, and Clear Vision (described in chapter 5), and these clearly should have been reported.[2]

Was this intentional or a case of one hand not knowing what the other was up to? That's not clear. Were Jefferson and kindred projects actual violations to the BWC? Also open to debate. Nevertheless, the projects were of grave concern to biological weapons nonproliferation experts. Even if the U.S. claim that the projects were defensive is accurate, they risked doing serious damage to the BWC. One European official observed the projects are "going to make it much easier for others to claim that work they are doing is legitimate biodefense work."[3] Michael Crowley, then of the British American Security Information Council, added pointedly: "If the U.S. administration had seen such work underway in other countries, then it would be the first to point the finger that this is questionable. And what this does is makes the grey areas greyer still between offense and defense, and that doesn't help."[4]

But it leaves other national governments two steps back, with no way of knowing what their counterparts are up to. They have to trust without verifying, a poor formula in a dangerous world.

Another Step Forward

In 1995, representatives of more than fifty BWC member nations began pulling together with the think tanks and institutes collectively called nongovernmental organizations, or NGOs, along with arms control advocates and lawyers, to give this gentleman's agreement the muscle of international law. The tool crafted over several years by the so-named Ad Hoc Group carried a title only a diplomat could love: Protocol to the Convention on the Prohibition of the Development, Production, and Stockpiling of Bacteriological (Biological) and Toxin Weapons and on Their Destruction.[5]

Only the word "protocol" was new, but its framers' goal was at every step to be clear, concise, and essential: to create a workable means of assuring compliance with the BWC. The path was as tortuous as any requiring consensus among so many diverse nations, but by 1998, the Ad Hoc Group had crafted the key elements of a compliance protocol and was hopeful that it would become international law. Then came more years of intensive meetings, negotiations, and discussions to work out details and find compromises. In the end, this enormous effort crafted by diplomats and technical experts from so many nations crashed and burned. The reasons are hugely complex. More important to the future, however, is this: Were the objections overwhelming, or could they still be met in hashing out a new compliance protocol?

There are two questions remaining in this book, and we believe they are among the most important questions we can ask. The first:

Had the protocol joined its Biological Weapons Convention as international law, would the world today be more dangerous or less dangerous?

The World of the Protocol

First, here is what would *not* have changed had the protocol been passed, even in its strongest form: al-Qaeda fanatics would have brought on the tragedy of 9/11, using the simple, relatively low-tech means that they did. The anthrax letters in all likelihood would have killed five people, sickened dozens more, and instilled fear throughout the populace, since all evidence points to the powder having been stolen from a U.S. facility.

But this would now be different: every member nation would have to put its dual-use cards face up on the table, a first step toward reducing suspicions and building confidence in each other's activities.

Here is the scenario that might have been:

The protocol requires every manufacturing plant, laboratory, or other facility with dual-use equipment above a certain size to declare it openly. Equipment that could be used to test or make biological weapons includes milling machines, aerosol generators, and large-scale fermentation tanks to brew microorganisms in large quantities.

The key is to ensure compliance with the BWC. This is accomplished by two different forms of international team inspections, called investigations and visits, and although the terms don't sound far apart, they will be conducted in very different atmospheres. "Facility investigations" can cover a number of situations. A facility investigation is launched by suspicion that a site is not complying with the BWC because it is developing, producing, or stockpiling bioweapons. Less seriously, one nation might suspect that another's undeclared lab or factory is actually dual use and simply should have been declared. After discussions with the nation's government, if suspicions still remain, a facility investigation could also be requested.

But at the most serious end of the spectrum, suspicion that a country is actually using bioweapons or detection of an unintentional release would launch a "field investigation." Investigations are accusatory by their nature, so launching them requires a vote of the executive council of those party to the protocol.[6]

But let's consider a contrasting situation where no wrongdoing is suspected—say, a manufacturing site that has already declared its dual-use equipment. This is where "transparency visits" come in, a key compliance force for the BWC. Declared sites around the world are randomly picked for transparency visits. No stigma here—the team is visiting this plant because its number came up. When the visit ends successfully, the firm may well put out a press release boasting of its role in keeping the world safe from biological weapons by allowing visits and bolstering its reputation for honest work. The atmosphere is friendly and the inspectors want to disrupt as little as possible.

The key difference between the two forms of oversight is that transparency visits simply aim to confirm compliance with the protocol's declaration requirements, and facility investigations look into potential violations of the BWC itself.

Transparency visits come with only a one-or two-day notice, so there is little time to cover up illegal activities. Site managers and their national governments alike would have major concerns that violations would be discovered if they never knew when transparency visitors were about to pay what is intended as a friendly call. And that, of course, offers reassurance for every nation under the protocol.

Douglas MacEachin, former deputy director of intelligence for the CIA, called the combination of facility investigations and transparency visits the twin pillars of the three supporting the protocol, the first being open declarations. As top intelligence officer in the CIA, the personable and extroverted MacEachin was highly expert in game-playing sleuthing scenarios. In arguing for the protocol's adoption he said, "The critical element that binds the on-site verification architecture together is that there is no treaty right-of-refusal for visits to declared sites and that those visits will be carried out in accordance with the agreed procedures to meet an agreed minimum level of transparency." But he warned, "This is an architecture within which the weakening or elimination of one pillar has a major impact on the remaining pillar."[7]

Now consider another major stumbling block: How do biodefense facilities, pharmaceutical companies, and university laboratories protect classified and confidential information during any inspection, friendly or not? Here, the word "inspection" signifies either a visit or and investigation. The Ad Hoc Group negotiators got major assistance from an unlikely ally, the U.S. Chemical Manufacturers Association,[8] the trade group for inspections under the sister Chemical Weapons Convention.[9] We call the association's alliance unlikely only because its member firms so often are owned by the same multinational corporations as those to be covered under the protocol. However, the origins of Managed Access lie in the chemical industry's efforts to improve its image in the wake of intense public criticism for developing defoliants and other chemicals used in Vietnam, as well as mounting international concern over chemical pollution.

The CWC already has built into it the three pillars of the BWC protocol: declarations, challenge investigations, and routine visits—the latter two analogs to facility investigations and transparency visits in the protocol. One way to look at the protocol is that it would have put the BWC on a parallel enforcement footing as that already in place for chemicals under the CWC.

Here is how Managed Access works.[10] An inspection team shows up at Major Chemical Corp. on a visit. The team wants to see what is being made inside Building A. But managers object that this will reveal a trade secret, a clever, very confidential—because it is very valuable—manufacturing process that goes on in Building A, so they ask to cover a critical piece of equipment with a tarp and that chemical engineers on the inspection team not be allowed in the building.

Tarp in place, they escort a nontechnical team member through the building from end to end, finishing where the product is dried and packaged. The team member observes that no special safety precautions are being taken in the building, so the process and final product are not dangerous. He or she might even take a sample of product right off the production line since this is to be shipped to customers. That accomplishes the team's goals while protecting the valuable process.

In Managed Access, the nation's government can always make the final decision over what must remain confidential. However, that government must try to find alternate means to satisfy inspectors' concerns and answer their questions. This is the key to Managed Access: Responsibility for transparency is still on the nation.

Now let's push a step further. What if, despite Managed Access being signed, sealed, and delivered by treaty, the host country stonewalls the inspectors? That certainly would arouse suspicions, and those suspicions could lead to a more intensive facility investigation. More stonewalling? Such strong indications of violation of the BWC might lead to economic sanctions, loss of standing in the international community, and other reprisals. Such threats are a deterrent, not a guarantee, but should make a nation's leaders think twice about violating the BWC.

Despite some problems along the way, Managed Access has worked, aiding international compliance with the Chemical Weapons Convention. John Gilbert, a senior science fellow at the Center for Arms Control and Non-Proliferation who has spent years involved firsthand in CWC compliance, says that after more than three thousand industry inspections worldwide, there is "high confidence that the inspection process has generally provided the information inspectors feel they need and that [Managed Access] has contributed to that perceived success."[11]

Gilbert adds that there have been bumps in the road, most early on and most eventually overcome. Those potential problems usually involved plant personnel or government representatives "not fully understanding the importance of demonstrating compliance," but were resolved when they realized that "taking a few minutes to cover some gauges before allowing inspectors to examine equipment" and taking other commonsense actions could eliminate problems. Significantly, he is unaware of any issue involving site operators or management feeling "they inadvertently gave up confidential information and suffered damage as a result."[12]

With such a major industry trade group as the Chemical Manufacturers Association actively on board the CWC, what went wrong with the protocol to the sister convention, covering the biological arena that is so much the focus of suspicion and fear?

The immediate answer is simple to tell, baffling to understand. All the varying treaty drafts, called rolling texts, and supporting papers submitted to the Ad Hoc Group conveyed the combined wisdom of more than fifty nations as well as their dozens of experts, ready for its chairman, Hungarian ambassador Tibor Toth, to finalize. Watch carefully: The Bush administration insisted on language that weakened the protocol—and that done, the United States then rejected the protocol for being weak.

As you might expect, that sleight of hand conceals an enormously complex scenario, whose major players ranged from large pharmaceutical companies to critics within the Bush administration. It was John Bolton, as undersecretary for arms control and international security at the U.S. State Department, who actually sabotaged the protocol in 2001. Without U.S. participation, which he refused, the protocol could not move forward towards an official vote.

Afterward Bolton said, "A country that is willing to lie about its compliance is a country that would be perfectly willing to sign the protocol and then lie about it.... We don't accept that it would be better than nothing. I would say it is worse than nothing.... It was our conclusion that more negotiations would probably not have solved it but made it worse."[13]

Back, now, to MacEachin of the CIA. He had addressed precisely this issue three years earlier, and did it so completely that it should be recalled in detail:

> Some opponents of a rigorous regime for non-challenge [transparency] visits argue that the nature of biological weapons programs is such that this concealment is easily done. Maybe. But how much confidence is the violator to have that this can be done? To what extent is the violator prepared to stake a weapons program on this gamble?[14]

He noted, "One argument has been that it takes only a short time after the departure of the visiting team to convert a legitimate biological research facility to production of biological weapons, citing physician Alan Zelikoff's January 8, 1998, *Washington Post* commentary. This is puzzling since it bypasses the issue of covering up all indications that a program was underway *before* the visiting team arrived."[15]

MacEachin went on to discuss two separate scenarios that would put the violator at risk of exposure and punishment. "If the cover-up takes place at a facility at which there are otherwise legitimate biological programs are all of the personnel working on the legitimate activities privy to the conspiracy? If not isn't there a risk that the cover-up in anticipation of a non-challenge visit could be detected by citizens who might leak the information further?"

"Indeed," he concluded, "experience has shown that often it is the cover-up efforts that expose the illicit activity, rather than the illicit activity itself. All things considered, these are risks that a regime seeking biological weapons probably would wish to avoid if possible."[16]

However, a major detractors' argument is that bioweapons agents can be destroyed in a short time, unlike hundreds of drums of chemicals agents. So, they assert, the BWC is unverifiable even with the added protocol, and no parallel can be drawn with verification under the CWC.

We agree that bioweapons agents might well be destroyed more readily, but we would argue that evidence of it would sometimes be left behind. Violators would always have the worry that destruction left telltale signs. Far from perfect, certainly, but still a considerable deterrent.

Being caught red-handed in a breach of the BWC could have carried serious consequences for a nation violating the longstanding agreement. It's hard to imagine how today's Russia could carry out the massive offensive bioweapons program undertaken by the Soviet regime, or that any other country could do so under the protocol, given the threat of a sudden appearance by friendly but savvy inspectors on a random transparency visit.

Bottom line: *If the protocol were in place we would be living in a safer world.*

While it is unlikely transparency visits would catch violators of the BWC red-handed—indeed, some would not even be suspected during a visit—it would uncover some of them, and that should leave would-be violators in a cloud of doubt. The combination would certainly contribute to a safer world. That seems eminently clear to us, for with the door to multilateral cooperation wide open, other steps could more easily be taken to increase our biosecurity by overcoming the protocol's flaws and strengthening its provisions. And let's not forget that the many billions of dollars spent in a fruitless effort to cure paranoia by defending against what villainy others might be up to could be put to far better uses.

The twin pillars of transparency visits and facility investigations would not guarantee compliance, but they would provide a deterrent to a nation or group of individuals contemplating a violation. Put yourself in the position of this plant manager: Your government has "suggested" that you secretly use one of your antibiotics fermenters to produce biological weapons. What kind of cover stories do you provide your employees? Do you ask them to knowingly lie or do you keep them in the dark? If they lie, will they convince inspectors who will be trained in criminal interrogation? What if even one employee cannot live with this in good conscience or simply gets rattled? On the other hand, if you try to keep skilled technicians and scientists in the dark, isn't it likely they will find you out?

Going forward, you will have to cover up all traces of your biological weapons work on little notice. Taken together, wouldn't you try to convince your government that the project is too risky to stake the nation's reputation—not to mention your own—on whatever doubtful benefits the work would bring?

Instead of covering up illegal activities in a *declared* facility, your country could build a site that you would operate wholly in secret, but then you are open to discovery by spy satellites, on-the-ground intelligence agents, or the brave or disaffected whistle-blower.

Most important of all, in the end, what do you or your nation hope to gain in a world where nearly every manner of lethal weapon is either *legally available* or easy to come by without detection? As we have seen, even acts of terror can be carried out far more readily with fertilizer and other chemicals bought over the counter.

Two Steps Back ... to Reality

All that, of course, would be the scenario in a world where a compliance protocol helped make the BWC enforceable. In the world we live in now, any of the dangerous actions we've just discussed would carry little or no risk of discovery, especially if carried out at the only level that is truly threatening—the rogue nation with sufficient capabilities.

Earlier we said there were two questions remaining in this book. Here is the second:

What's stopping us now?

We said that behind the "too strong/too weak" sleight of hand in America's wrecking the protocol lay a complex of actors and motives. Were any of the roadblocks thrown up insurmountable?

The major counterforce was the blow struck by the U.S. government, which feared that its biodefense secrets would be given away. In an open and transparent world, most national endeavors to develop a biosecurity apparatus should be everyone's business, because they ought to protect everyone. The sort of dangerous knowledge that ought to be secret would remain so. An example: finding a way to infiltrate biosecurity goals by understanding how to make microbes resistant to new antibiotic countermeasures should and would remain a closely guarded secret.

But something else was at work in the U.S. opposition that we have seen overtly or implied throughout this book: insistence on going it alone, mounting purely national initiatives and regulations, and pressuring other nations to do the same. That attitude must give way to an understanding that only multinational efforts can succeed.

The second major opponent of the protocol was the pharmaceutical industry, which perennially sees itself as overregulated, overinspected and, most important, exquisitely vulnerable to loss of its core business by loss or theft of nothing more than a single microbe culture. Unlike even their sibling chemical giants, pharmaceutical companies profit from their multimillion-dollar, decades-long investments on the powers of single strains of carefully nurtured, antibiotic-producing bacteria. Their adage is "a single bacterium is the factory." Despite these concerns, industry opposition in Europe was tamer and when opposition appeared it was mild. Much of Europe's opposition appeared to be the result of pressure from U.S. industry.

The industry feared intentional theft of "the factory" by an unscrupulous inspector. But as we've seen, that legitimate concern can be addressed, as under Managed Access, in laying out ground rules at the beginning of a visit. Inspectors can be required to wear company-supplied sterile gowns, which they would leave behind when they shower at the end of the tour, and they would be accompanied by the firm's representatives from start to finish.[17]

A second fear is more complex. Investigations and visits are conducted under international agreement whose decision makers are the inspected national governments, not site managers. So might inspections leave pharmaceutical companies at the mercy of political power plays? The government would have the authority to override a company's pleas for protection of its valuable secrets, and it conceivably might do so if politically expedient.[18] That seems a legitimate worry when a company's fortunes might hang in the balance.

Here, however, the United States might enable agreement with industry rather than skewer it. The U.S. government certainly would have the power to stand behind its companies, and other national governments would most likely follow its lead. That assurance should be given.

Let's continue answering that second vital question, What is stopping us now? First and foremost, the international political will to move forward. Under a new administration, the United States could lead the international community to regain that will. And we will need both the will and the support to succeed. There is a major block in the road that could derail the protocol if we do not work together to remove it.

Enter Catch X

The next roadblock seems the ultimate deal breaker. A vocal group of developing nations made a demand that had to be heard, for it followed language already in the BWC. The group insisted on "the right to participate in ... the fullest possible exchange of equipment, materials, and scientific and technological information for the use of bacteriological (biological) agents and toxins for peaceful purposes."[19]

The statement calling for "fullest possible exchange" is a direct quotation from Article X of the Biological Weapons Convention. In other words, all countries are *entitled* to all this, they pointed out, and before they would sign the protocol, they wanted a guarantee that that provision of the BWC finally would be carried out. What could be more reasonable?

This: for the safety of all the world, keeping biological weapons and the means to develop them out of the hands of rogue nations and what are termed "nonstate actors"[20]—that is, terrorists—must be the overarching goal. The Australia Group of nations [21] adamantly opposed export of dualuse equipment and materials to certain states, some of which would have been protocol members.

So the problem stands, to this day, as a classic dilemma in which two inherently contradictory principles appear to have no common resolution.

But this problem has a resolution. Carrying it out will not be simple, but at the end of the road to a new protocol, the world will be a safer place. Here is what our solution would look like:

> Dual-use equipment should be made available to all parties of the BWC and its new protocol. The dual-use items would be indelibly embossed with identification numbers. To receive equipment under Article X, recipients would need to describe exactly where and how the equipment would be used, and they would have to agree to special visits to those sites for confirmation. In turn, the Australia Group would need to change its policy to allow transfer of dual-use items to any country that is party to both the BWC and the protocol. Visits always were integral to the protocol, so an equipment-inspection requirement, perhaps during transparency visits, would add no extraordinary complications or demands.

Virulent pathogens, including bioweapons strains, must be strictly controlled. In short, transfer of dangerous pathogens, which may be interpreted as allowed in article X for peaceful purposes, should be off the table in protocol discussions and dealt with elsewhere. This will help simplify new protocol negotiations by removing one contentious issue. A future international agreement on controlling dangerous pathogens must be reached under any circumstances, and we will consider one we think lays solid groundwork for doing so.

Finally, declarations, visits, and investigations of pharmaceutical facilities should be reinserted into the protocol.[22] Alternatively, in place of visits, the industry could provide tours of their facilities, which we describe in the epilogue.

Through the Back Door—Almost

The protocol would have established an organization similar to the Organization for the Prohibition of Chemical Weapons, which was established for the Chemical Weapons Convention. The OPCW has a number of regulatory powers, including the important ability to conduct on-site inspections of industrial facilities utilizing chemicals that could be used in the manufacture of chemical weapons.

A few years ago, the parties to the BWC made a modest start toward creating a related organization; the Implementation Support Unit was established in 2006. With only three employees, the ISU is indeed modest, and it has little authority.

The Hudson Institute's Richard Weitz observed that "the ISU remains primarily an international support center rather than a policy making or regulatory agency." He added, "Unlike the OPCW and the International Atomic Energy Agency (IAEA), the ISU has no mandate to enforce compliance or even monitor whether the ... parties are submitting accurate information about their activities."[23]

What, then, does the ISU do? It organizes meetings and assists BWC parties in carrying out confidence-building measure obligations and in writing laws to enforce the BWC in their own countries. The unit also seeks to enroll new members in the Biological Weapons Convention.

At the BWC's 2007 yearly meeting, several countries urged expansion of the tiny unit, and the European Union was prepared to finance it. But the United States, sensing that this was a back-door attempt toward establishing a verification regime for the BWC, again used its clout, this time to thwart expansion, as it had earlier to thwart establishing the protocol. There things stood as the Bush administration came to an end and the Obama administration took the reigns in January 2009.

A New Day?

The new administration now is putting its stamp on major policies both domestic and international. If, as we hope, bringing a measure of true biosecurity to the world is among its top priorities, elected officials and administrators will find waiting for them the already-built foundations for a new compliance protocol and other means of enforcing the Biological Weapons Convention.

Many NGOs and university groups, supported by foundations or by grants, have worked tirelessly to anticipate the contingencies of a more complex world of biological developments and international relations. Like the rolling texts and papers given Chairman Toth in the original protocol discussions, these represent the collected wisdom of many of the world's best minds on the subject of global biosecurity. Here are a few we find most promising.

THE BIOSECURITY TRUST

One person who, early on, both recognized the need for a multifaceted approach to critical biosecurity issues and articulated how such an approach might work was Rob Sprinkle, a physician with a PhD in international affairs. Sprinkle named his web of security measures the Biosecurity Trust.[24] He suggested global standards for research safety as well as for research responsibility among corporations and other institutions.

Sprinkle called for a range of methods to track potentially high-risk science. Some are formal, such as a system for actually classifying extraordinarily provocative research results, others less so, such as a worldwide biosecurity "confessional box" for whistle-blowers. Still others could be forged from resources at hand—say, databases to organize public information like the career paths of life scientists in "more worrisome states, laboratories, and corporations," and those in "sub-fields with the clearest potential for abuse."[25] Whether influenced by the Sprinkle article or not, many of his suggestions are now being implemented or are under consideration around the world.

THE HARVARD SUSSEX PROPOSAL

Harvard's Matthew Meselson and Julian Robinson, a senior fellow at the University of Sussex, drafted a treaty that would make development of biological or chemical weapons an international crime.[26] Their treaty would join that outlawing air piracy—hijacking aircraft—in requiring all supporting countries to prosecute or agree to extradition of anyone violating its provisions.

Their likening the criminalization of bioweaponry to air piracy provides a powerful counter to those who claim that international sanctions would not work because some countries would flout them. It was only in 1970 that the Hague Convention [27] imposed criminal sanctions on airline hijackers and required any nation in which they sought refuge to either prosecute them or allow their extradition. Before that, such events were regularly in the headlines. Now they are rare, even though a few countries have violated the treaty and granted hijackers asylum. Clearly, the agreement radically cut the odds of succeeding and just as sharply increased the penalty for failure. It may not be perfect, but it works.

Closer to home, consider legal speed limits. Few would argue that they are unnecessary, even though they are flouted by most of us at some time and by some of us every day. Far from perfect, but they work.

Meselson and Robinson note that neither the BWC nor the CWC contain such criminal provisions: "Treaties defining international crimes are based on the concept that certain crimes are particularly dangerous or abhorrent to all." Certainly, they say, the hostile use of disease or poison or turning the gifts of biotechnology into biological weapons would threaten not only every nation, but future generations of every nation. Their proposed convention "would make it a crime under international law for any person knowingly to develop, produce, acquire, retain, transfer, or use biological or chemical weapons or knowingly to order, direct or render substantial assistance to those activities or to threaten to use biological or chemical weapons."[28]

And the enforcer: "Any person who commits any of the prohibited acts anywhere would face the risk of prosecution or extradition should that person be found in the territory of a state that supports the proposed convention."[29]

SECURITY AND OVERSIGHT OF PATHOGENS AND TOXINS

An international treaty would eliminate the bewildering tangle of national laws governing security for pathogenic microorganisms and toxins in favor of a single "rule book," under a proposal by Michael Barletta, Amy Sands, and Jonathan Tucker. In 2002 they argued that most commerce in pathogens and select agents is perfectly legitimate but international, so only international law can competently track shipments and set safety standards.[30]

And that commerce is stunning in scope as well as in the mishmash of rules now governing it. A prime example: "There are 46 germ banks—in countries as diverse as Germany, India, and Iran—that stock anthrax cultures," they pointed out, citing a report by the World Federation for Culture Collections, "a loose association of 472 repositories of living microbial specimens in 61 countries."

The federation dutifully urged members to tighten access to dangerous microbes. Urge is all it can do. But even if the federation could enforce such obviously vital strictures, there are still more than a thousand germ banks worldwide that are not members, and the authors say, "few of their culture collections are adequately secured or regulated."

In addition to other public health threats posed by such a lack of uniform regulation, combating a new epidemic or pandemic would require rapid international cooperation—a difficult feat at best that could fail under the burden of the different sets of licenses and documentation each country now requires.

PROTOTYPE FOR PROTECTIVE OVERSIGHT

Earlier we discussed in some detail the comprehensive "Biological Research Security System" developed at the University of Maryland's Center for International and Security Studies.[31] One element of this model is particularly useful in considering the varied responsibilities for oversight in innumerable laboratories and other facilities conducting experiments around the world. The group proposes that for work at less dangerous levels, local oversight would suffice. But work judged of extreme concern—a few activities that, say, might result in pathogens "significantly more dangerous than currently exist"—would trigger international oversight. Here are a few of our candidates for the "extreme concern" red flag:

- Working with virulent 1918 pandemic flu virus or SARS.
- Experimenting on the cowpox virus's mechanisms for modulating its host's immune system.
- Any work on virulent smallpox virus.
- Trying to make avian flu contagious in humans in order to understand how that could happen naturally.

By now, our reasons for putting this work under widest, tightest scrutiny should be plain: escape from the lab in every case could spark a pandemic.

The current situation: There is *no* international proactive oversight of these or any other dangerous experiments—that is, no means to prevent ill-conceived or unnecessary experiments from being undertaken, let alone for monitoring them.

The BWC Has Us Covered

Alan Pearson argues that some of the legal foundations for domestic oversight already are in hand, from two sources. Article IV of the BWC requires that signers must ban and prevent any private action that would defeat the convention's objectives. And a UN resolution similarly requires members to enact and enforce laws prohibiting any nonstate actor—terrorists, companies, anyone—from pursuing biological weapons development.[32]

But what about national governments themselves? Pearson says the BWC mandates the same Article IV rules for states, their subunits and their agents, a point made by Germany at a compliance review meeting some years ago.[33]

Pearson's clever attempt to bring in sorely needed oversight of dangerous activities through the back door is commendable, but likely will fail to gain support. Why? As we've seen over and over, the U.S. delegation under Bush resisted. Time will tell if the Obama administration adopts a new stance.

"You're either part of the solution or you're part of the problem."

Black revolutionary Eldridge Cleaver's aphorism,[34] often quoted but rarely attributed, seems especially appropriate for what we urge of everyone, with no insurgency overtones needed.

Which of the strategies we've outlined should be adopted? All of them, and others still to come, as long as they are cost effective and don't have a clear downside that might ultimately decrease our biosecurity or interfere with legitimate science. No single strategy, however sound, is even close to being perfect, but taken together all of them will significantly increase biosecurity and therefore be part of the solution.

The United States tends to view biosecurity narrowly, as defense against a biological weapons attack and pandemic flu. We must understand biosecurity broadly, as protection of people, animals, and plants from all deliberate, accidental, *or natural* infectious diseases and chemical or biological toxins. Our approaches to biosecurity must then be equally broad.[35] Achieving it is a job for everyone from scientists, who must be aware of their ethical responsibilities, to nations, which must observe and enforce the Biological Weapons Convention, to citizens, whose active support is a requisite for good new ideas to take root and come to life.

Good biodefense policy is only one element of a sane program to cut risks of hostile or dangerous activities. With the mushrooming of BSL-*3* and BSL-*4* laboratories not only here but around the world, dangerous activities and access to bioweapons agents will increase dramatically, all the more so if nations of the world, no matter how hostile to one another, fail to reach some form of consensus on this issue. By definition, every experiment carried out in these labs is dangerous or it would be done at lower—and less costly—containment level.

High-biosecurity labs are expensive to build and maintain, so their existence must be justified, and we fear that unnecessary dangerous experiments will be devised simply to justify their existence, marking yet another ratcheting up of the hazards that will come at us from all the directions we have been discussing. Here is yet another compelling reason for winning international protection for whistle-blowers—albeit a long shot, judging from recent headlines here and abroad, since any "biosecurity confessional box" would place loyalty to higher ethical codes above loyalty to corporate or government superiors.

Working scientists must take leading roles to prevent and combat hostile exploitation and dangerous activities in biology. Recruiting them would be a prime requisite for developing defensive or offensive biological weapons programs, as it was everywhere in the past. A number of strategies are being discussed to make scientists more aware of their ethical responsibilities and more able to carry them out under what certainly would be trying circumstances. And as we noted earlier, scientists one day may have their own version of the Hippocratic oath, committing them first to "do no harm."

For all the scarcities facing the world right now, there is no shortage of bright and able people thinking hard on the problems we have discussed here. They come from every field and bring an enormous range of expertise—and, clearly, stamina.

The Biosecurity Trinity

From all we and others have urged throughout, three common concepts repeatedly emerge as a trinity essential to attaining biosecurity: transparency, cooperation, oversight.

"Trinity" seems an appropriate term to conclude this hard look into the dismal past and hopeful future of biosecurity. A sacred religious symbol meant to herald peace, it also names the site of the first nuclear explosion that foreshadowed how interdependent the security of all nations would become. That was the first "genie out of the bottle," just a few years before the elaboration of the structure of life's master molecule released the many genies of the new biology.

Paranoia and permissiveness are the surefire symptoms of disease, and the integrated force of transparency, international collaboration, and oversight is the cure—and inoculation against future outbreaks.

As the world's most powerful nation, the United States will inevitably lead the way, for better or for worse. Before we can accomplish anything, however, we must regain the moral high ground by changing our way of acting in the world. In the biological and chemical weapons arenas this means supporting the strongest interpretations of both the BWC and the CWC and embracing the biosecurity trinity. Saber rattling and bullying along any front of international activity will sow distrust on all fronts. We have a special obligation to lead in moral behavior.

If thinking globally does not lead to acting globally, we will face a future as dangerous as our paranoia has conjured. We will have made it so. And we must not, either actively or passively.

It is dangerous to America and the world that these calls to action are largely unheard by the public—and even by much of the science community. Those who have read this book have already listened. We hope the word will spread. There are vital roles for everyone, and there are no sidelines.

Epilogue

President Barack Obama's wholehearted support for scientific research and his early pursuit of international cooperation on several fronts offer real prospects for a sound biological security policy. So far, the administration's focus has rightly been on nuclear weapons. For the foreseeable future, the odds are far greater of mass deaths caused by a nuclear device than by a biological weapon.

However, now is the time to back off the massive funding of dangerous experiments that pose our greatest risk of lethal disease outbreaks from accident or theft. And now is the time to reengage in international efforts to protect against biological warfare by truly strengthening the Biological Weapons Convention—not simply offering lip service to it.

Consider the advice offered to the president by a highly diverse group of biosecurity experts regarding the need for international cooperation. The group said that the U.S. government "has paid inadequate attention to prevention and response measures internationally, thereby increasing our vulnerability to a significant biological event and heightening the skepticism of other countries about our commitment to either improving global public health or reducing deliberate and accidental biological risks to global security."[1]

What is remarkable about this group's agreeing that the government has increased our vulnerability to "biological events" and heightened world skepticism about our intentions is its composition. Convened by the Center for Arms Control and Non-Proliferation, the group is made up of members who represent the entire spectrum of expert opinion we have been discussing. Some, for example, strongly supported the protocol to the Biological Weapons Convention; some equally vigorously opposed it.

As the new administration took the world stage, other experts weighed in. Matthew Meselson, whose advice played such a key role during the Nixon administration, suggested that active oversight and transparency begin here at home, urging that Congress devise national guidelines and procedures covering all government activities, ranging from conducting compliance reviews to assuring reliability of laboratory workers and others engaged in biodefense.[2]

Meselson didn't stop there. He said federal law should make provisions for site visits, require periodic reporting on projects, and mandate observers from states and the U.S. Justice Department, who, among other duties, would provide expertise in U.S. law and treaty commitments. The goal, he said, is to create a model for future international efforts.

Among those encouraging and admonishing the incoming president were, not surprisingly, some harking back to past policies and fears. Among the most prominent, unfortunately, was Dr. Tara O'Toole, the CEO of the Center for Biosecurity at the University of Pittsburgh Medical Center.

She and the center's Thomas Inglesby wrote that "there are no technical barriers that prevent state programs, non state groups [terrorists], or individuals from building and using a biological weapon that could sicken or kill as many as tens of thousands of people or more, and that could lead to grave societal disruption and economic damage."[3]

We hope by now we have established that a terrorist attack using biological weapons may be likely, but one with severe, massive consequences is extraordinarily unlikely. If the fearmongering about mass deaths seems over the top, it should come as no surprise that O'Toole is the author of the over-the-top-by-design Dark Winter and TOPOFF exercises, in which smallpox and plague attacks, respectively, quickly overwhelmed the entire health care system. By design: the numbers were fed into the programs' scenarios.

Why do we call O'Toole's appearance on the stage unfortunate? Because soon after making that statement, she was named Department of Homeland Security undersecretary for science and technology. Her appointment drew wide-ranging criticism from biosecurity experts for just the sort of hyperbole she demonstrated both in the scenarios and the statement.

On the plus side, appointed secretary of Homeland Security was Janet Napolitano, who, as governor of Arizona, was highly critical of TOPOFF and whose statements on national security issues since her appointment have favored international cooperation.

We believe others of the president's key appointments also bode well for the future. As of this writing, the twenty members of the President's Council of Advisors on Science and Technology include molecular biologists Eric Lander, renowned for his work in human genomics, and Harold Varmus, a Nobel laureate and former NIH director; and astrophysicist Christopher Chyba.[4] The biosecurity policies that Chyba has promoted closely parallel those for which we have been arguing.[5]

Margaret Hamburg, now head of the FDA, has frequently supported the alarmist views we oppose. However, she has had a long, distinguished career in public health and policy.[6] Given that today's major public health issues involve emerging infectious diseases, antibiotic resistance in pathogens, and the threat of pandemics, to us she is the right person at the right time to head the agency charged with drug approval.

President Obama seems assured a steady stream of advice that we consider sound on ways to strengthen the BWC and reopen discussions that might finally bring openness and transparency to international biosecurity efforts.

We will turn to the world's pharmaceutical companies, which might take their own steps toward international transparency independently of any government initiative. From our perspective, significant international transparency regarding compliance to the BWC cannot be achieved without some sort of onsite activity. Here, the drug industry can help jumpstart the international transparency process at almost no cost to itself.

During talks for a compliance protocol to the BWC, PhRMA, the trade association representing more than forty large U.S. drug companies, offered a scenario that might be played out against an American firm. The scenario highlighted companies' deepest concerns about possible consequences of the protocol.

Since the protocol would have been an agreement among nations, the U.S. government would have had the legal right to make decisions regarding an international visiting team's access to confidential information, materials, and sensitive sites within a drug facility. However, PhRMA argued, only facility managers would have the knowledge to make such decisions.[7] That notwithstanding, the manager's decision might be overridden by, say, the government's political desire to answer all the visiting team's questions. That is, for political expediency, an official might allow access despite the consequences, turning the firm into a sacrificed pawn in international diplomacy.

We would suggest that the world's drug companies revive an idea presented by an industry representative during informal protocol discussions. In place of "transparency visits," the companies themselves would offer bona fide international inspection teams "tours" of their production facilities. The difference: the company, not the government, would have complete control over what visitors could see, and the company would approve who was on the touring team. That would guarantee the security of vital business information. In this manner, the pharmaceutical industry clearly would boost its international image, sullied by its stubbornness during the protocol talks, and show itself willing to help reduce the threat of biological weapons development.

What would the world gain? Only an agreement allowing some form of international onsite visits can build confidence among nations that others are not developing biological weapons. Government biodefense facilities—not industrial plants—obviously would be the critical targets of such visits. A move by the industry to allow at least some form of visitation could remove it from the discussion, allowing negotiations to focus on governments' willingness to open their facilities to one another.

This year also gave the world an unwelcome, but we think critically important, look into multinational response to a global pandemic. The H1N1 swine flu outbreak may be receding into history as you read this, pushed aside by new urgencies. Or it may be *the* new urgency, the relatively non-virulent virus having mutated. There is no way of knowing now, in May 2009, what will have occurred by the end of the year or beyond. But in our view, the global response by officials and experts has offered a solid precedent for international cooperation and openness in all aspects of public health.

The outbreak's epicenter was in Mexico. Mexican health officials discovered it, quickly reported it to the CDC and WHO, and took aggressive measures to limit its spread. Initially, there were several highly publicized scare statements by some in the United States, but the discussion was quickly modulated by criticism from others. Debate about the likelihood and severity of a possible resurgence in the fall flu season continues, and it should. Had the strain been extremely lethal as well as contagious, it would have taken more than a coordinated response to control a global pandemic, but just such a coordinated response would have been all the more essential. We should consider all potential biological threats in the same light.

For some of the most deadly potential bioweapons—smallpox would be the prime example—the ultra-rapid spread of disease without regard for borders means that no state or terrorist organization could unleash one without destroying its own foundations. Nevertheless, however obvious this may be to forward-looking political leaders and public citizens, unforeseen events of precisely the type brought on by terrorists can at a stroke undermine public support for cooperative policies and harden already-defensive attitudes into isolationism and paranoia.

Terrorism aims to disrupt social order to the greatest degree possible; the less effort needed to do so, the better. That suggests we put a new twist on Franklin D. Roosevelt's maxim: the primary weapon of terrorism we need to fear is fear itself. Fear is the fuel of paranoia, and paranoia is a disease of isolation. Its natural counterforce is carefully reasoned cooperation and openness, a mutual openness with visible guarantees of integrity.

The story of Messina with which we began offers a parable. If the Caffa siege survivors' ship was the vector for the enemy's biological weapon to make landfall in Europe, the natural-born plague soon smothered much of Asia, Europe, and North Africa, and that in the days when nothing could be communicated faster than on horseback or under sail. Today it takes a single aircraft, merchant vessel, international train, or truck a few hours to do far worse. Would the disease vector be a terrorist or innocent disease victim? Whatever the answer, how would the answer alter what our immediate response must be?

Notes

Chapter One

[1] Barbara W. Tuchman, *A Distant Mirror: The Calamitous Fourteenth Century* (New York: Alfred A Knopf, 1978). Tuchman cites several scholarly sources supporting the Messina outbreak's origin among Caffa survivors. Some scholars, such as Mark Wheelis, argue that the Caffa refugees likely spread their infection to Constantinople, the Messina epidemic then arising as other ships arrived from Constantinople. Mark Wheelis, "Biological Warfare at the 1346 Siege of Caffa," *Emerging Infectious Diseases* 8, no.9 (Sept. 2002): 971–75. Also, a lay-level multimedia presentation that the reader may find entertaining on the medieval plague epidemics and related history may be found in the online course developed by Lynn C. Klotz et al., "Biosecurity: Risks, Responses, and Responsibilities (Washington, DC: Center for Arms Control and Non-Proliferation), placed online August 2005. http://www.politicsandthelifesciences.org/Biosecurity_course_folder/base.html (accessed Feb. 27, 2009).

[2] Mark Wheelis, "Biological Warfare before 1914," chapter 2 in *Biological and Toxin Weapons: Research, Development and Use from the Middle Ages to 1945,* ed. Erhard Geissler and John Ellis van Courtland Moon, Chemical & Biological Warfare Studies No. 18 (Stockholm: SIPRI,

1999). Smallpox was used as a biological weapon against Native Americans in the 18th century. During the French and Indian War (1754–1767), Sir Jeffrey Amherst, commander of British forces in North America, suggested the deliberate use of smallpox to "reduce" Native American tribes hostile to the British. Without knowledge of Amherst's suggestion, on June 24, 1763, Captain Ecuyer, the fort's commander, gave blankets and a handkerchief from the smallpox hospital to the Native Americans. A trader at the fort, William Trent, recorded in his journal, "I hope it will have the desired effect." While this adaptation of the Trojan horse ruse was followed by epidemic smallpox among Native American tribes in the Ohio River valley, other contacts between colonists and Native Americans may have contributed to these epidemics. Smallpox epidemics among immunologically naive tribes of Native Americans following initial contacts with Europeans had been occurring for more than 200years. In addition, the transmission of smallpox by fomites was inefficient compared with respiratory droplet transmission.

[3] Mark Wheelis, "Biological Sabotage in World War I," in *Biological and Toxin Weapons: Research, Development and Use from the Middle Ages to 1945.* Further explanation is provided by Martin Furmanski: an account in a "post–WW I 'memoir'

of a German spy talks of receiving a report of hundreds of dead horses on a ship going to Mesopotamia, but British Veterinary Service records record no such event. I looked through the records of the U.K. Army Veterinary Service for WW I, both in Europe and in the U.S. procurement areas. Anthrax was of course well known and easily recognized and diagnosed, and there were only 2 outbreaks of anthrax with 3 horses dead during the entire war in France, and these occurred in areas known to harbor anthrax in the soil. Glanders was also well known and was the subject of very intense surveillance ... one must realize that the 769 equids lost to glanders was truly trivial, considering that the U.K. army maintained about 300,000 horses 'on active duty' on the Western Front. As far as the French horses are concerned, although the French caught a German saboteur with glanders cultures in 1917, if he caused many glanders cases it did not affect their war effort either" (e-mail message to Lynn Klotz, Sept. 12, 2007).

[4] For convenience we use a single number—forty million for deaths from the 1918 pandemic flu. Sources report a wide range of numbers for deaths. For example, "Review of 1918 Pandemic Flu Studies Offers More Questions than Answers" on the National Institute for Allergy and Infectious Diseases Web site reports between fifty to

one hundred million deaths (NIH press release, Feb. 28, 2007; http://www3.niaid.nih.gov/news/newsreleases/2007/1918PressReleaseASF.htm [accessed Feb. 27, 2009]). Discussion on the Stanford University Web site reports between twenty and forty million deaths worldwide ("The Influenza Pandemic of 1918," http://www.stanford.edu/group/virus/uda/ [accessed Feb. 27, 2009]; "Pandemics and Pandemic Threats since 1900" reports more than fifty million deaths, http://www.pandemicflu.gov/general/historicaloverview.html [accessed Feb. 27, 2009]).

[5] The full report on which Rhodes's testimony was based is "High-Containment Biosafety Laboratories: Preliminary Observations on the Oversight of the Proliferation of BSL-3 and BSL-4 Laboratories in the United States," statement of Keith Rhodes, Chief Technologist Center for Technology and Engineering Applied Research and Methods, at the Hearing of House Energy and Commerce Committee Subcommittee on Oversight and Investigations, Oct. 4, 2007, http://www.gao.gov/new.items/d08108t.pdf.

[6] Mark Williams, "The Knowledge Part 3: The Current Revolution in Biotechnology Is More Likely to be Exploited by National Militaries than by Terrorists," *Technology Review,* March 15, 2006, http://www.technologyreview.com/BioTe

ch/wtr_16584,312, p1.html (accessed Feb. 27, 2009).

[7] Milton Leitenberg, James Leonard, and Richard Spertzel. "Biodefense Crossing the Line," *Politics and the Life Sciences* 22, no.2 (2003): 2–3.

[8] http://www.usamriid.army.mil/aboutpage.htm.

[9] Convention on the Prohibition of the Development, Production, and Stockpiling of Bacteriological (Biological) and Toxin Weapons and on their Destruction, April 10, 1972. See http://www.opbw.org/convention/documents/btwctext.pdf.

[10] President Richard Nixon, "Remarks Announcing Decisions on Chemical and Biological Defense Policies and Programs," speech delivered on Nov. 25, 1969. Cited in John T. Woolley and Gerhard Peters, The American Presidency Project (Santa Barbara, CA: University of California [hosted], Gerhard Peters [database]), http://www.presidency.ucsb.edu/ws/index.php?pid=2344 (accessed Feb. 27, 2009). The first half of the quote may also be found in a statement issued by Nixon: *Foreign Relations, 1969–1976,* Vol. E-2, *Documents on Arms Control, 1969–1972* (released by the Office of the Historian), 166. Statement Issued by President Nixon, Nov. 25, 1969, http://www.state.gov/r/pa/ho/frus/nixon/e2/83597.htm (accessed Feb. 27, 2009).

[11] http://www.bartleby.com/73/1257.html. The quotation has been attributed to Khrushchev many times, as the Web site notes in its citations, but has never been verified in any of Khrushchev's writings or speech transcriptions.

Chapter Two

[1] "Report of the Defense Science Board Task Force Future Strategic Strike Forces," Office of the Undersecretary of Defense for Acquisition, Technology and Logistics, Washington DC, 20301-3140 Feb. 2004, Section 7 page 18 (emphasis added), http://www.acq.osd.mil/dsb/reports/fssf.pdf(accessed March 3, 2009).

[2] Our book will focus mainly on incapacitating agents, not the full range of neuroweapons and military research in neurobiology. An account of the full range of neuroweapons may be found in the book by bioethicist Jonathan Moreno, *Mind Wars: Brain Research and National Defense* (New York: Dana Press, 2006).

[3] Matthew Meselson is Thomas Dudley Cabot Professor in the Department of Molecular and Cellular Biology of the Natural Sciences at Harvard University. He is also Faculty Chair for CBW Studies at the Belfer Center for Science and International Affairs at the John F. Kennedy School of Government.

[4] Matthew Meselson, "Averting the Hostile Exploitation Of Biotechnology," *CWC Conventions Bulletin* 48 (2000): 16–19.

[5] "Appeal on Biotechnology, Weapons and Humanity," International Committee of the Red Cross. The press release and full text may be found at http://www.icrc.org/web/eng/siteeng0.nsf/htmlall/5eamtt?opendocument (accessed March 11, 2009).

[6] Robin Coupland, e-mail message to Lynn Klotz.

[7] A. Cohen, "Staph Bacteria—Most Common Cause of Hospital-Based Infection—Cost New York City $435.5 Million and Claimed 1,400 Lives in 1995," Eureka Alert!" http://www.eurekalert.org/pub_releases/1997-06/RU-SBCC-120697.php (accessed March 11, 2009)

[8] R.A. Howe, N.M. Brown, R.C. Spencer, "The New Threats of Gram Positive Pathogens: Re-Emergence of Things Past," *Clinical Pathology* 49 (1996): 444–49.

[9] Ken Alibek with Stephen Handleman, *Biohazard,* (New York: Dell Publishing, 1999), 160.

[10] Variola virus, complete genome, dsDNA; linear; length: 185,578 nucleotides; created: May 4, 1993, http://www.ncbi.nlm.nih.gov/entrez/viewer.fcgi?db=nucleotide&val=X69198.

[11] J. Cello, A.V. Paul, E. Wimmer, "Chemical Synthesis of Poliovirus cDNA: Generation of Infectious Virus in the Absence of Natural

Template," *Science* 297 no.5583 (Aug. 9, 2002): 1016–18. http://www.sciencemag.org/cgi/content/full/297/5583/1016 (accessed March 11, 2009).

[12] Smallpox presents an additional complication not present in polio virus. With polio virus, simply introducing the virus's genome into a cell will produce an infectious virus. For smallpox, additional "helper" proteins are required to produce an infectious virus. While there are straightforward strategies for supplying these proteins, it is unclear whether the strategies will work given the present state of our knowledge.

[13] J.B. Tucker and C. Hooper, "Protein Engineering: Security Implications," *EMBO reports* 7 (July 1, 2006): S14. http://www.nature.com/embor/journal/v7/n1s/pdf/7400677.pdf (accessed March 3, 2009).

[14] Ibid.

[15] Ibid.

[16] "Newly Discovered Small Molecules," *Scripps Research Institute, EurekAlert!* March 14, 2006, http://www.eurekalert.org/pub_releases/2006-03/sri-nds031406.php.

[17] J.A. Di Masi, R.W. Hansen, H.G. Grabowski, „The Price of Innovation: New Estimates of Drug Development Costs," *Journal of Health Economics* 22, no.2 (March 2006): 151–85.

[18] "Profile 2008: PhRMA" Pharmaceutical Research and Manufacturers of America, 2008, http://www.phrma.org/files/2008%20Profile.pdf (accessed March 3, 2009).

[19] In 1962, James D. Watson shared the Nobel Prize with Francis Crick in Physiology or Medicine for elucidating the structure of DNA nine years earlier.

[20] David A. Wheeler et al., "The Complete Genome of an Individual by Massively Parallel DNA Sequencing," *Nature* 452 (April 17, 2008): 872–76.

[21] As reported in the English online edition: "Russia Warily Eyes Human Samples," *Kommersant,* May 30, 2007, http://www.kommersant.com/p769777/r_527Human_medical_biological_materials_export (accessed March 3, 2009). This was also reported in the United States in an editorial: Barry Kellman, "The Potential Dark Side of Genetics," *The San Francisco Chronicle,* July 8, 2007, http://www.sfgate.com/cgi-bin/article.cgi?file=/c/a/2007/07/08/INGAHQQ8DN1.DTL (accessed March 3, 2009).

[22] An alternative explanation was provided by Cyrill Vatomsky: "The general idea was to try to cut into the smuggling of human organs business but the only way Federal Customs and Russian officialdom in general know how to react is to ban everything

outright." (http://cyrillvatomsky.com/index.cfm/2007/5/30/Why-in-the-world-would-Russia-ban-exports-ofbiological-specimen.

[23] The American International Health Alliance was concerned enough that it published a denial that they use Russian samples for genetic weapons development. The AIHA establishes and manages partnerships between health care institutions in the United States and their counterparts in central and Eastern Europe (http://www.aiha.com/en/NewsAndEvents/Press Releases/2007/PR_2007-06-01.asp [accessed March 3, 2009]).

[24] When we speak of nonlethal agents here, we refer only to chemical and biological agents that would be used to alter the mental and physical states of targeted individuals. We are not referring to other types of nonlethal weapons, such as beanbag guns, Tasers, etc.

[25] William J. Broad, "Oh, What a Lovely War. If No One Dies," *New YorkTimes,* Nov. 3, 2002. Despite its eye-catching title, the article examines both the pros and cons of nonlethal weapons.

[26] The Russian health minister publicly identified the gas as fentanyl shortly after the attack, according to the *New York Times* and other news organizations (Michael Wines and Sabrina Tavernise, "Russia Recaptures Theater after Chechen Rebel Group Begins to Execute

Hostages," Oct. 26, 2002). However, some still argue whether other gases were involved and say the matter is unresolved. See Boris Nemtsov, "Putin Places Popularity over People," *St. Petersburg Times,* Oct. 30, 2007. Nemtsov is a siege negotiator and member of the federal political committee of The Union of Right Forces, a political party in the State Duma.

[27] Clem Cecil, "Chechen Siege Hostages Still Dying Of Gas Effects," *London Times,* May 1, 2003.

[28] Sharon Weinberger, "Pentagon's Psychic Vision Revisited," *Wired.com,* June 28, 2007, http://blog.wired.com/defense/2007/06/dinner-with.html(accessed March 3, 2009).

[29] The thought that the Chechens were likely going to shoot all the hostages may be erroneous. According to Martin Furmanski, "It's now quite clear that the Russian Govt *[sic]* initiated the chemical attack without immediate provocation, and that the Chechens were actively negotiating and showing no signs of shooting hostages .... They intentionally killed two individuals who voluntarily entered the theater after the hostages were taken. One (a woman) entered the theater and tried to exhort the theater hostages to make an open revolt and rush the terrorists, and the other was a man who claimed to have a son hostage in the theater, but could not find him. Both were considered

spys *[sic]* and shot" (e-mail message to Lynn Klotz, Feb. 2008).

[30] Lynn Klotz, Martin Furmanski, and Mark Wheelis, "Beware the Siren's Song: Why 'Non-Lethal' Incapacitating Chemical Agents are Lethal," issued paper from the Center for Arms Control and Non-Proliferation, March 2003 http://fas.org/bwc/papers/sirens_song.pdf(accessed March 4, 2009).

[31] Till Manzke, Ulf Guenther, Evgeni G. Ponimaskin, et al., "5-HT4(a) Receptors Avert Opioid-Induced Breathing Depression Without Loss of Analgesia," *Science* 301 (July 11, 2003): 226–29.

[32] E.G. Ponimaskin, M.F. Schmidt, M. Heine, U. Bickmeyer, D.W. Richter, "5-Hydroxytryptamine 4(a) Receptor Expressed in Sf9 Cells Is Palmitoylated in an Agonist-Dependent Manner," *Biochemical Journal* 353, no.3 (2001): 627–34.

[33] Neil Davison, "'Off the Rocker' and 'On the Floor': The Continued Development of Biochemical Incapacitating Weapons," Bradford Science and Technology Report No. 8, Bradford Disarmament Research Centre (BDRC), Department of Peace Studies, University of Bradford, Aug. 2007, http://www.brad.ac.uk/acad/nlw/research_reports/docs/BDRC_ST_Report_No_8.pdf (accessed March 4, 2009).

[34] Ibid., 19.

[35] Ibid., 37–39.

[36] Julian Robinson and Matthew Meselson, "Non-Lethal Weapons, the CWC and the BWC,", *CBW Conventions Bulletin* 61 (Sept. 2003): 1–2 (original emphasis).

[37] "Nonlethal Weapons and Capabilities," Report of an Independent Task Force Sponsored by the Council on Foreign Relations, 2004,http://www.cfr.org/search.html ?q=calmatives&ie=&site=cfr&output=xml_no_dtd&client=cfr&lr=&proxystylesheet=cfr &oe=&getfields=authors. pubtype&x=17&y=9.

[38] "The Use of Drugs as Weapons: The Concerns and Responsibilities of Healthcare Professionals," British Medical Association, May 2007, http://www.bmj.com/cgi/content/extract/334/7603/1073-a (accessed March 4, 2009).

[39] U.S. Congress, The Select Committee to Study Governmental Operations with Respect to Intelligence Activities, Foreign and Military Intelligence, Church Committee Report no.94–755, 94th Cong., 2d Sess. (Washington, DC: Government Printing Office, 1977), 394.

[40] U.S., Senate, Select Committee on Intelligence, and Subcommittee on Health and Scientific Research of the Committee on Human Resources, Aug. 3, 1977.

[41] The effects of BZ "include disorientation with visual and auditory hallucinations. The agent disturbs the higher integrative functions of memory, problem solving, attention, and com-

prehension" (Tom J. Evans, "Chemical Warfare," *Kirk-Othmer Encyclopedia of Chemical Technology,*http://mrw.interscience.wiley.com.ezproxy1.lib.asu.edu/emrw/9780471238966/kirk/article/chemharr.a01/current/pdf [accessed March 22, 2009]).

[42] James S. Ketchum, *Chemical Warfare: Secrets Almost Forgotten, A Personal Story of Medical Testing of Army Volunteers with Incapacitating Chemical Agents During the Cold War (1955–1975)* (Santa Rosa, CA: ChemBooks Inc., 2006).

[43] Ibid., 33.

[44] Ibid., 128.

[45] Michael Kosfeld, Markus Heinrichs, Paul J. Zak, et al., "Oxytocin Increases Trust in Humans," *Nature* 435 (June 2, 2005): 673–76, http://www.nature.com/nature/journal/v435/n7042/abs/nature03701.html (accessed March 4, 2009).

[46] http://www.verolabs.com/salestool.php?UID=2009030410242475.69.142.20 (accessed March 7, 2009).

[47] Antonio R. Damasio is Director of the Brain and Creativity Institute at the University of Southern California and the author of several bestselling books dealing with the brain, emotions, and self.

[48] Antonio Damasio, "Brain Trust," *Nature* 435 (June 2, 2005): 571.

[49] Moreno, *Mind Wars,* 8.

[50] Emily Singer, "The Dangers of Synthetic Biology: Nobel Prize Winner David Baltimore Explains Why Building Smallpox from Scratch Is a Key Safety Concern in Synthetic Biology," *Technology Review,* May 30, 2006, http://www.technologyreview.com/Biotech/16932/.

[51] Moreno, *Mind Wars,* 12. There is a small error in this quote: The Internet was first called ARPANET, not DARPAnet, as the agency was not yet called "DARPA."

Chapter Three

[1] For an excellent, detailed account of the forces at work and history leading up to Nixon's decision, see Jeanne Guillemin, *Biological Weapons: From the Invention of State-Sponsored Programs to Contemporary Bioterrorism* (New York: Columbia University Press, 2005), esp. chapter 6.

[2] Nixon, "Remarks Announcing Decisions on Chemical and Biological Defense Policies and Programs" (see ch.1, note 10).

[3] Editor's note, *Foreign Relations, 1969–1976,* Vol. E-2, *Documents on Arms Control, 1969–1972* (released by the United States Office of the Historian).

[4] Lynn Klotz interview with Matthew Meselson, May 2006.

[5] Ibid.

[6] Ibid.

[7] April 1965. Matthew Meselson, "A Proposal to Inhibit the Development of Biological Weapons," *Proceedings of the Fourteenth Pugwash Conference on Science and World Affairs* (April 1965): 297–304, http://www.mindcontrolforums.com/matthew-meselson.htm.

[8] Matthew Meselson, book review of *Tomorrow's Weapons, Chemical and Biological* by Jacquard Hirshon Rothschild, *Bulletin of the Atomic Scientists* (Oct. 1964): 35–36, http://www.mindcontrolforums.com/matthewmeselson.htm

[9] Ibid.

[10] Interview with Matthew Meselson, May 2006.

[11] Interview with James Leonard, Sept. 2007.

[12] Second interview with James Leonard, Oct. 2007.

[13] Interview with James Leonard, Sept. 2007.

[14] Ibid.

[15] Second interview with James Leonard, Oct. 2007.

[16] "1925 Geneva Protocol," United Nations Office for Disarmament Affairs, http://www.un.org/disarmament/WMD/Bio/pdf/Status_Protocol.pdf.

[17] Reservations, such as this one, are not unusual and even normal in international treaties.

[18] Interview with James Leonard, Sept. 2007.

[19] There was considerable pressure on nations on both sides of World War II to use chemical weapons. For an analysis as to why, in the

end, they were not used, see Julian P. Robinson and Milton Leitenberg, "The Non-Use of CB Weapons during World War II," chapter 5 in *The Problem of Chemical and Biological Weapons,* SIPRI Yearbook (Almqvist & Wiksell, 1971).

[20] Edward Cody, "War Lives On at Museum of the Macabre," *Washington Post,* April 7, 2006, http://www.washingtonpost.com/wp-dyn/content/article/2006/04/06/AR2006040602022_pf.html.

[21] Jeanne Guillemin, "Imperial Japan's Germ Warfare: The Suppression of Evidence at the Tokyo War Crimes Trial, 1946–1948," chapter 8 in *Terrorism, War, or Disease? Unraveling the Use of Biological Weapons,* ed. Anne L. Clunan, Peter R. Lavoy, and Susan B. Martin (Palo Alto, CA: Stanford University Press, 2008).

[22] SFE 188/2, State-War-Navy Coordinating Committee for the Far East, August 1, 1947, RG 153, MFB, U.S. National Archives. Quoted in Guillemin, "Imperial Japan's Germ Warfare," 171, 177.

[23] Martin Furmanski, e-mail message to Lynn Klotz, Sept. 12, 2007.

[24] Undocumented reports indicate that the Russians may have used biological weapons in World War II against the invading Germans and in Manchuria against the Japanese as well.

[25] The source for much of the information on the Japanese biological weapons activities in Manchuria is Sheldon Harris's scholarly study of the Japanese biological weapons program, *Factories of Death: Japanese Biological Warfare, 1932–1945, and the American cover-up* (New York: Routledge, 2002), esp. the preface and chapters 1–3; also conversations via e-mail and in person with Martin Furmanski, who had served as pathologist in Manchuria during Sheldon Harris's investigation and helped Harris during the preparation of the second edition of his book.
[26] Furmanski, e-mail message to Lynn Klotz.
[27] Harris, *Factories of Death,* 102
[28] Ibid.
[29] Ibid.
[30] Ibid., 101
[31] Furmanski, e-mail message to Lynn Klotz.
[32] Klotz et al., "Biosecurity: Risks, Responses, and Responsibilities (see ch.1, note 1), unit 2.
[33] Interview with James Leonard, Sept. 2007.
[34] Harris, *Factories of Death,* 56–57.
[35] Ibid., 57 (emphasis added).
[36] Testimony of Dr. Schalk van Rensberg, Truth and Reconciliation Commission Hearing into Chemical and Biological Warfare, part 4, Cape Town South Africa, June 9, 1998, http://www.doj.gov.za/trc/special/index.htm#cbw.

[37] Testimony of Dr. Schalk van Rensberg, Truth and Reconciliation Commission Hearing into Chemical and Biological Warfare, part 4, Cape Town, South Africa, June–July 1998, http://www.doj.gov.za/trc/special/cbw/cbw4.htm.

[38] Ibid.

[39] Ibid.

[40] Ibid.

[41] Chandré Gould, pers. comm. with Lynn Klotz and Ed Sylvester, 2006.

[42] Van Rensburg's assertion that Yellow Rain was a Soviet chemical weapon likely comes from Vietnam-era U.S. intelligence beliefs, subsequently challenged by a group of U.S. scientists led by Matthew Meselson. See Matthew Meselson, Jeanne Guillemin, Julian Robinson, "Yellow Rain: The Story Collapses," *Foreign Policy,* Fall 1987, 100–17. See also Jonathan Tucker, "Conflicting Evidence Revives Yellow Rain Controversy," John Martin Center for Nonproliferation Studies, http://cns.miis.edu/stories/020805.htm.

[43] Guillemin, Jeanne. "Seduced by the State," *Bulletin of the Atomic Scientists* 63, no.5 (2007): 14–16.

[44] Interview with James Leonard, Sept. 2007

[45] Alibek and Handleman, *Biohazard,* 40–41 (see ch.2, note 9).

[46] Interview with James Leonard.

[47] Kanatjan Alibekov was born in Kauchuk, Kazakhstan, in 1950. His has a PhD in microbiology and an MD in military medicine. After defecting to the United States in 1992, he Americanized his name to Ken Alibek.

[48] Eric Croddy and Sarka Krcalova: "Tularemia, Biological Warfare, and the Battle for Stalingrad (1942–1943), *Military Medicine* 166, no.10 (Oct. 2001): 837–38, http://cns.miis.edu/research/cbw/tula.htm.

[49] Alibek and Handelman, *Biohazard,* 102.

[50] Ibid., prologue, x.

[51] Ibid., 101.

Chapter Four

[1] Kyle B. Olson, "Aum Shinrikyo: Once and Future Threat?" *Emerging Infectious Diseases* 5, special issue (July–Aug. 1999): 513–16. All sources seem to agree that seven people were killed. However, the Monterey Institute of International Studies reports that 144 confirmed serious injuries and 126 more people complaining of symptoms such as headaches, vision impairment, nausea and vomiting, etc. (Tim Ballard, Jason Pate, Gary Ackerman, and Sean Lawson, "Chronology of Aum Shinrikyo Chemical, Biological, and Related Incidents," March 13 2001, http://cns.miis.edu/pubs/reports/um_chrn.htm). aAn anonymous reviewer of our book says that 58 were sent to the hospital

and 253 to clinics. The exact numbers are not important for our book.

[2] Ballard et al., "Chronology of Aum Shinrikyo Chemical, Biological and Related Incidents." Other accounts claim that the attacks sickened thousands (Aum Bankruptcy Case Wraps Up," *Asahi Shimbun,* Nov. 27, 2008, http://www.asahi.com/english/Herald-asahi/TKY200811270083.html).

[3] Agence France Presse, "Psychiatrist Backs Japan's Doomsday Guru," June 30, 2006, http://www.nationmultimedia.com/worldhotnews/readphp?newsid=30007617..

[4] "Joyu Tells Authorities His Group to Leave Aum," *Yomiuri Shimbun,* March 6, 2007.

[5] Olson, "Aum Shinrikyo."

[6] Holly Fletcher, "Aum Shinrikyo (Japan, cultists, Aleph, Aum Supreme Truth)," Council on Foreign Relations, http://www.cfr.org/publication/9238/#4.

[7] The Center for Defense Information, "Aum Supreme Truth (Aum) Aum Shinrikyo," May 23, 2002. http://www.cdi.org/friendlyversion/printversion.cfm?documentID=879.

[8] U.S. Centers for Disease Control and Prevention, "Q Fever," http://www.cdc.gov/ncidod/dvrd/qfever/index.htm#overview1.

[9] Olson, "Aum Shinrikyo."

[10] Paul Keim et al., "Molecular Investigation of the Aum Shinrikyo Anthrax Release in

Kameido, Japan," *Journal of Clinical Microbiology* 39, no.12 (Dec. 2001): 4566–67. Vaccines are often prepared from harmless or attenuated strains of the disease microorganism. The theory is that these strains, without making us too sick, will produce the immune-system memory needed for future protection from the disease.

[11] U.S. Centers for Disease Control and Prevention, "Ricin," http://emergency.cdc.gov/agent/ricin/.

[12] There were five recovered anthrax letters and three presumed anthrax letters that were not recovered. William Robert Johnson, "Review of Fall 2001 Anthrax Bioattacks," U.S. Centers for Disease Control and Prevention, http://www.cdc.gov/niosh/nas/RDRP/appendices/chapter6/a6–45.pdf.

[13] "Update: Investigation of Bioterrorism-Related Anthrax and Adverse Events from Antimicrobial Prophylaxis," *Morbidity and Mortality Weekly Report* 50, no.44 (Nov. 9, 2001): 973–76, http://www.cdc.gov/mmwr/preview/mmwrhtml/mm5044a1.htm.

[14] U.S. Environmental Protection Agency, "Danbury, Conn. Anthrax Clean Up,", press release, Sept. 2007, http://epa.gov/region1/er/sites/danbury/.

[15] Commission for the Prevention of Weapons of Mass Destruction Proliferation and Terrorism,

Bob Graham, Jim Talent, *World at Risk: The Report of the Commission on the Prevention of Weapons of Mass Destruction Proliferation and Terrorism* (New York: Vintage Books, 2008), 8.

[16] U.S. Centers for Disease Control and Prevention, "Bioterrorism Agents/Diseases" Emergency Preparedness and Response, http://www.bt.cdc.gov/agent/agentlist-category.asp#catdef. We have modified and slightly expanded the CDC list.

[17] Thomas V. Inglesby, Donald A. Henderson, John G. Bartlett, et al., "Anthrax as a Biological Weapon," *JAMA* 281 (1999): 1735–45.

[18] Douglas Beecher, "Forensic Application of Microbiological Culture Analysis to Identify Mail," *Applied and Molecular Microbiology* 72, no.8 (Aug. 2006): 5304–10.

[19] The case numbers in this article refer to the listing of cases in "American Anthrax Outbreak of 2001," UCLA Department of Epidemiology, School of Public Health, http://www.ph.ucla.edu/epi/bioter/detect/antdetect_list.html. A thorough and readable account of the anthrax letter attack and its victims up to 2003 is offered in Leonard Cole, *The Anthrax Letters: A Medical Detective Story* (Washington, DC: Joseph Henry Press, 2003).

[20] Scott Shane, "Anthrax Survivors Find Life a Struggle," *Baltimore Sun*, Sept. 18, 2003.

[21] Milton Leitenberg and Mark Wheelis, pers. comm.
[22] The major reference for this section is D.A. Henderson, T.V. Inglesby, Bartlett J.G., et al., "Smallpox as a Biological Weapon," *JAMA* 281, no.22 (June 9, 1999): 2127–37.
[23] U.S. Centers for Disease Control and Prevention, "Smallpox Fact Sheet," Emergency Preparedness and Response, http://emergency.cdc.gov/agent/smallpox/overview/disease-facts.asp.
[24] Martin Furmanski, e-mail message to author, Sept. 28, 2007.
[25] Henderson et al., "Smallpox as a Biological Weapon," 2127.
[26] Furmanski, e-mail message to author.
[27] Unless otherwise noted, the reference for this section is Thomas V. Inglesby; David T. Dennis; Donald A. Henderson, et al., "Plague as a Biological Weapon," *JAMA* 283 (2000): 2281–90.
[28] Ibid., 2282.
[29] "Seventh Arrest in Ricin Case," BBC News, Jan. 8, 2003, http://news.bbc.co.uk/2/hi/uk_news/2637515.stm.
[30] "Florida Man Faces Bioweapons Charge," CNN, Jan. 14, 2005, http://www.cnn.com/2005/US/01/13/ricin.arrest/index.html.
[31] Ashley Powers, "Man with Ricin Is Sentenced; Roger Bergendorff Said He Never Planned to

Use the Homemade Toxin that Authorities Found in Las Vegas," *Los Angeles Times,* Nov. 18, 2008.

[32] These definitions follow those proposed by S. Bennet, Risk Assessment Program Manager, Committee on Methodological Improvement to the Department of Homeland Security's 2006 Bioterrorism Risk Assessment, in an unpublished PowerPoint Presentation entitled "DHS Bioterrorism Risk Assessment: Background, Requirements, and Overview" (Aug. 28, 2006), http://www.armscontrolcenter.org/assets/ppt/bennett_dhs_bioterrorism_082806.ppt#649,9, Overview of the 2006 DHS Bioterrorism Risk Assessment Approach.

[33] Most of the material in this section is drawn from David T. Dennis; Thomas V. Inglesby; Donald A. Henderson, "Tularemia as a Biological Weapon," *JAMA* 285 (2001): 2763–73.

[34] Richard Preston, *The Hot Zone* (New York: Anchor Books, 1994).

[35] J.H. Kuhn, "Marburgviruses and Ebolaviruses—History, Fiction, and the Facts" (seminar, MIT Faculty Dinner Series on Biosecurity, Sept. 29, 2005).

[36] E.K. Leffel and D.R. Reed, "Marburg and Ebola Viruses as Aerosol Threats," *Biosecurity and Bioterrorism: Biodefense Strategy, Practice, and Science,* vol.2, no.3 (2004), 186–91.

[37] J. H Kuhn, e-mail message to Lynn Klotz, Jan. 2008.

[38] Kuhn, "Marburgviruses and Ebolaviruses."

[39] Ibid.

[40] U.S. Centers for Disease Control and Prevention, Special Pathogens Branch, "Filoviruses," http://www.cdc.gov/ncidod/dvrd/spb/mnpages/dispages/filoviruses.htm.

[41] Leffel and Reed, "Marburg and Ebola Viruses as Aerosol Threats."

[42] Furmanski, e-mail message to Lynn Klotz.

[43] UNSCOM, "Major Sites Associated With Iraq's Past WMD Programs," Global Security.org , Dec. 3, 1997, http://www.globalsecurity.org/wmd/library/news/iraq/un/971203_sites.htm .

[44] David Kay, Carnegie Endowment for International Peace/Century Foundation/Georgetown University Forum, "Combating Weapons of Mass Destruction," Feb. 5, 2004. Kay's talk may be found at http://www.leadingtowar.com/claims_sources/2004.02.05%20kay%20carnegie.doc (accessed March 23, 2009).

[45] *New York Daily News,* Nov. 17, 1997.

[46] Guillemin, *Biological Weapons,* 161 (see ch.3, note 1).

[47] Maureen Dowd, "Anthrax, Shmanthrax," *New York Times,* Nov. 19, 1997.

[48] The most recent (2007) U.S. census estimate for the population of Washington, D.C., was

588, 292. U.S. Census Bureau, State and County Quick Facts, http://quickfacts.census.gov/qfd/states/11000.html(accessed March 20, 2009).

[49] Fernando L. Benitez and Larissa I. Velez-Daubon, "CBRNE—Nerve Agents, V-series: Ve, Vg, Vm, Vx," *eMedicine Journal* 3, no.1(Jan. 11, 2002).

[50] Mesleson, "A Proposal to Inhibit the Development of Biological Weapons" (see ch.3, note 7). This is his now-famous Pugwash paper where he compared nuclear, chemical, and bioweapons as part of his strategic arguments against bioweapons development. Meselson used metric units to cite 1,000 kilograms, or 1 metric ton, as the lethal sarin dose per square kilometer (1,000/170=5.88 kg of VX per sq km).

[51] The area of Washington, D.C., is 61.40 square miles, according to the U.S. Census Bureau, State and County Quick Facts, http://quickfacts.census.gov/qfd/states/11000.html (accessed March 20, 2009).

[52] Ketchum, *Chemical Warfare,* 262 (see ch.2, note 42).

[53] Ibid.

[54] Each warhead would cover a circle of area=π (450)[2]=635,850 sq ft, so to cover the required area of 1,964 million square feet, 1,544 warheads would be needed.

[55] Mark Wheelis, "A Short History of Biological Warfare and Weapons," in *The Implementation of Legally Binding Measures to Strengthen the Biological and Toxin Weapons Convention,* ed. M.I. Chevrier, K. Chomiczewski, M.R. Dando, H. Garrigue, G. Granaztoi, and G.S. Pearson (Amsterdam: ISO Press, 2003), 15–31.

[56] U.S. Department of Justice, Federal Bureau of Investigation, "Terrorism 2002–2005," http://www.fbi.gov/publications/terror/terrorism2002_2005.htm.

[57] See, for example, Katarina Kratovac, Associated Press, "Iraq Chlorine Attacks Raise New Concerns," Feb. 22, 2007.

Chapter Five

[1] The agency formulating the relevant policy under the act is the National Institute of Allergy and Infectious Diseases, a division of NIH.

[2] Center for Arms Control and Non-Proliferation, "Federal Funding for Bioweapons Prevention and Defense" (2009), http://www.armscontrolcenter.org/policy/biochem/articles/fy09_biodefense_funding/.

[3] Alibek and Handelman, *Biohazard,* 53 (see ch.2, note 9).

[4] Chandré Gould and Peter I. Folb, "The Role of Professionals in the South African Chemical and Biological Warfare Programme," *Minerva* 40 (2002): 77–91.

[5] "Biodefense for the 21st Century" (White House press release, April 28, 2004).
[6] Ibid.
[7] W. Seth Carus, "Defining 'Weapons of Mass Destruction,'" Center for the Study of Weapons of Mass Destruction, National Defense University, Jan. 2006. Carus is deputy director of the Center, located at the National Defense University.
[8] Michael A. Levi and Henry C. Kelly, "Weapons of Mass Disruption," *Scientific American,* Nov. 2002,75–81.
[9] One biological weapon, anthrax, because of its long term stability in the environment, may make buildings uninhabitable for a long time, akin to needing to rebuild after destruction.
[10] Quoted in John Tierney, "Can Humanity Survive? Want to Bet on It?" *New York Times,* Jan. 30, 2007.
[11] Milton Leitenberg, "Bioterrorism, Hyped," *Los Angeles Times,* Feb. 17, 2006...
[12] Jack Melling, e-mail message to the Center for Arms Control and Non-Proliferation's Scientists Working Group, of which he is a member (used with permission).
[13] Quoted in Lawrence Wright, *The Looming Tower: Al-Qaeda and the Road to 9/11* (New York: Alfred A. Knopf, 2006), 303.
[14] Donald A. Henderson, Thomas V. Inglesby Jr., and Tara O'Toole, "A Plague on Your City:

Observations from TOPOFF," *Clinical Infectious Diseases* 32, no.3 (2001): 436–45 The authors of the papers describing the exercises TOPOFF and Dark Winter, as well as the participants, are from The Johns Hopkins Center for Civilian Biodefense Strategies, which is part of the University of Pittsburgh Medical Center.

[15] T. O'Toole, M. Mair, and T.V. Inglesby, "Shining Light on 'Dark Winter,'" *Clinical Infectious Diseases* 34 (2002): 972–83.

[16] Since this TOPOFF exercise there have been three more—TOPOFF 2, 3, and 4—that have led to the same conclusions about our lack of preparedness for a major bioweapons attack. Descriptions of these exercises may be found at http://www.fema.gov/news/newsrelease.fema? id=2806 (TOPOFF 2); http://www.njslom.org/TOPOFF_3_Exercise. pdf (TOPOFF 3); and http://www.dhs.gov/xprepresp/training/gc_1179430526487.shtm (TOPOFF 4).

[17] Bill Durodie, "Facing the Possibility of Bioterrorism," *Current Opinion in Biotechnology* 2004, no.15:264–68.

[18] Ballard et al., "Chronology of Aum Shinrikyo Chemical, Biological, and Related Incidents"(see ch.4, note 1).

[19] Alan Pearson, "Fiscal Year 2009 Federal Funding for Bioweapons Prevention and Defense," Center for Arms Control and Non-Proliferation, April 15, 2008, http://www.armscontrolcenter

.org/policy/biochem/articles/fy09_biodefense_funding/.

[20] As noted, the information came from the *Times* story (Miller, Engleberg, and Broad, "In Secretly Fighting Germ Warfare, US Tests Limits of a 1972 Treaty," *New York Times,* Sept. 4, 2001); however we also cite the BASIC report where the projects are summarized. Michael Crowley, "Disease by Design: De-Mystifying the Biological Weapons Debate," research report for the British American Security Information Council, Nov. 2001, http://www.basicint.org/pubs/Research/2001diseasebydesign1.htm .

[21] This list derives from a 2004 presentation by Lieutenant Colonel George Korch entitled "Leading Edge of Biodefense: The National Biodefense Analysis and Countermeasures Center" (http://www.cbwtransparency.org/archive/nbacc.pdf). It was made public when accidentally placed on the Internet.

[22] Reorganized from Department of Homeland Security, "Fact Sheet: National Biodefense Analysis and Countermeasures Center" (Feb. 2, 2005), http://www.dhs.gov/xnews/releases/press_release_0611.shtm.

[23] Private communication between a member of the Center for Arms Control and Non-Proliferation Scientists Working Group and the diplomat.

[24] Milton Leitenberg, James Leonard, and Richard Spertzel, "Biodefense Crossing the Line," *Politics and the Life Sciences* 22, no.2 (May 2004): 2–17.

[25] Anon., pers. comm. with Lynn Klotz. Understandably, the scientist does not want his name used as it would hurt his relationship with the Russian scientists.

[26] Joby Warrick, "The Secretive Fight Against Bioterror" *Washington Post,* July 30, 2006.

[27] Gerald L. Epstein, "Security Is More than Public Health," *Biosecurity and Bioterrorism: Biodefense Strategy, Practice, and Science* 5, no.4 (2007): 353–58 (quote is from p.355). The commentary was in response to Lynn C. Klotz, "Casting a Wider Net for Countermeasure R&D Funding Decisions," *Biosecurity and Bioterrorism: Biodefense Strategy, Practice, and Science* 5, no.4 (2007): 313–18.

[28] Quoted in Stephen Flynn, *The Edge of Disaster: Rebuilding a Resilient Nation,* (New York: Random House, in cooperation with the Council on Foreign Relations, 2007), 92–94.

[29] Lynn Klotz interview with David Ozonoff, Oct. 12, 2007.

[30] Ibid.

[31] "Within Our Grasp—Or Slipping Away? Assuring a New Era of Scientific and Medical Progress: A Statement by a Group of Concerned Universities and Research Institutions," sponsored by

Columbia University, The Johns Hopkins School of Medicine, The Johns Hopkins University, Partners Health Care, Washington University of St. Louis, University of Wisconsin, and Yale University (March 2007). The second report is "A Broken Pipeline? Flat Funding of the NIH Puts a Generation of Science at Risk," sponsored by Brown University, Duke University School of Medicine, Harvard University, Ohio State University Medical Center, Partners Healthcare, University of California–Los Angeles, and Vanderbilt University (March 2008). Both reports may be accessed at http://www.brokenpipeline.org.

[32] "A Broken Pipeline?" 1.

[33] See, e.g., http://www.super-science-fair-projects.com/biology-science-fairprojects.html ; http://www.swrp.org/Student_Presentations/Abstracts/columbriversandison_abstracts.html ; http://www.science-projects.com/biology.htm.

[34] "Biosafety in Microbiological and Biomedical Laboratories (BMBL)," 4th ed., BMBL sec. III: Laboratory Biosafety Level Criteria, http://www.cdc.gov/search.do?queryText=Laboratory+Biosafety+Level+Criteria&search Button.x=0&searchButton.y=0&action=search.

[35] Keith Rhodes, "High-Containment Biosafety Laboratories" (see ch.1, note 5).

[36] Scott Shane, "Army Suspends Germ Research at Maryland Lab," *New York Times,* Feb. 10,

2009. The 219 number is for labs registered with the CDC. Other labs have likely registered with the USDA, which could add significantly to the total number of labs working with anthrax.

[37] Reported in Nick Schwellenbach, "Biodefense: A Plague of Researchers," *Bulletin of the Atomic Scientists* 61, no.3 (May/June 2005): 14–16.

[38] As recalled by Lynn Klotz.

[39] William Frist, "Manhattan Project for the 21st Century," (Seidman Lecture, Harvard Medical School, Department of Health Care Policy, June 1, 2005).

[40] The Brookings Institution, "The Costs of the Manhattan Project (Expenditures through August 1945)," http://www.brookings.edu/projects/archive/nucweapons/manhattan.aspx. Cost in 1996 dollars was converted to 2006 dollars using yearly consumer-price-index inflation data (GPEC Information Center).

[41] Donald Kennedy, "Science and Secrecy," *Science* 289, no.5480 (2000): 724. Quoted in Julie Fischer, "Stewardship or Censorship? Balancing Biosecurity, the Public's Health, and the Benefits of Scientific Openness" (Washington, DC: Henry L. Stimson Center, 2006), 23–37.

[42] U.S. Centers for Disease Control and Prevention, Final Rule (42 CFR 73), http://www.selectagents.gov/. The actual list as of November

17, 2008, may be found at http://www.selectagents.gov/resources/List%20of%20 Select%20Agents%20and%20Toxins_111708.pdf.

[43] Ibid.

[44] The Public Health Security and Bioterrorism Preparedness and Response Act (2002), the Agricultural Bioterrorism Protection Act (2002), and the U.S.A. Patriot Act (2001).

[45] 42 CFR parts 72 and 73, *Federal Register* 70, no.152 (March 18, 2005), 13305 http://www.selectagents.gov/resources/42_cfr_73_final_rule.pdf.

[46] Fischer, "Stewardship or Censorship?" 23–27.

[47] Gary Splitter, quoted in Harvey Black, "The Test of Terrorism," *Milwaukee Journal Sentinel,* Jan. 8, 2006, http://www.jsonline.com/alive/news/jan06/383237.asp.

[48] Fischer, "Stewardship or censorship?" 24.

[49] Ibid., 25–26.

[50] Ibid., 25.

[51] Ibid., 25.

[52] http://www.ncbi.nlm.nih.gov/sites/entrez

[53] In a few cases, abstracts are not provided when you request them; instead, the following message appears "id: 16740818 Error occurred: Document retrieval error: document does not exist."

[54] Fischer, "Stewardship or censorship?" 29.

[55] Ibid., 30.

[56] Ibid., 30.

[57] Information on the Butler case is from George J. Annas, "Bioterror and Bioart—A Plague o' Both Your Houses," *New England Journal of Medicine* 354, no.25 (June 22, 2006): 2715–20.

[58] Many news accounts over the years have detailed the investigation and aftermath: Dan Hardy, "Chester Man Wants His Name Cleared in Anthrax Case," *Philadelphia Inquirer,* Aug. 10, 2008; William J. Broad et al., "For Suspects, Anthrax Case Had Big Costs," *New York Times,* Aug. 10, 2008; Dan K. Thomasson, "FBI Screwups Still Haunt Anthrax Investigation," Capitol Hill Blue, Oct. 10, 2006, http://www.capitolhillblue.com/content/2006/10/fbi_screwups_st.html; Dan Hardy et al., "FBI Raids Homes in Chester in Probe," *Philadelphia Inquirer,* Nov. 14, 2001.

[59] Hardy, "Chester Man Wants His Name Cleared in Anthrax Case."

[60] Lynne Duke, "The FBI's Art Attack Offbeat Materials at Professor's Home Set Off Bioterror Alarm," *Washington Post,* June 2, 2004.

[61] "On with the Show," editorial, *Nature* 429 (June 17, 2004): 685.

[62] Paula Reed Ward, "Scientist Gets Light Sentence for Mailing Bacteria," *Pittsburgh Post-Gazette,* Feb. 12, 2008.

[63] "Teacher from Buffalo Cleared of Bioterrorism," *Pittsburgh Tribune-Review,* April 23, 2008, ht

tp://www.pittsburghlive.com/x/pittsburghtrib/news/cityregion/s_563749.html?source=rss&feed=1.

[64] Ibid.

Chapter Six

[1] Perhaps the highest profile supporter of the al-Qaeda theory is Ross Getman, a lawyer from Syracuse, New York, and a prolific writer on the subject. Getman's Web site carries the title "Codename Zabadi: Ayman Zawahiri's Infiltration of US Biodefense" (http://www.anthraxandalqaeda.com/).

[2] Joby Warrick, "Trail of Odd Anthrax Cells Led FBI to Army Scientist," *Washington Pos* t, Oct. 27, 2008.

[3] Scott Shane, "Portrait Emerges of Anthrax Suspect's Troubled Life," *New York Times,* Jan. 4, 2009.

[4] Warrick., "Trail of Odd Anthrax Cells Led FBI to Army Scientist."

[5] Shane, "Portrait Emerges of Anthrax Suspect's Troubled Life."

[6] Larry Margasak, Associated Press, "U.S. Labs Mishandling Deadly Germs," Oct. 2, 2007.

[7] http://hosted.ap.org/specials/interactives/wdc/biohazards/. The Sunshine Project, "Texas A&M Bioweapons Accidents More the Norm than an Exception" (news release, July 3, 2007). See also Associated Press interactive map "Problems at U.S. Labs that Research Deadliest Biological

Agents (http://hosted.ap.org/specials/interactives/wdc/biohazards/).

[8] The scientific paper describing what has become known as the Australian Mousepox Experiment is: R.J. Jackson, A.J. Ramsay, C.D. Christensen, S. Beaton, D.F. Hall, and I.A. Ramshaw "Expression of Mouse Interleukin-4 by a Recombinant Ectromelia Virus Suppresses Cytolytic Lymphocyte Responses and Overcomes Genetic Resistance to Mousepox," *Journal of Virology* 75, no.3 (Feb. 2001): 1205–10.

[9] Recently, Ron Jackson, the lead author on "Expression of Mouse Interleukin-4" (ibid.), claimed that the experimenters never intended the IL-4 modified mousepox virus to be used in the field because of its relation to smallpox (http://www.fas.org/biosecurity/education/dualuse/FAS_Jackson/3_A.html).

[10] A discussion of the implications of the Australian Mousepox Experiment is presented in Federation of American Scientists, "Mousepox Case Study, Module 4," http://www.fas.org/biosecurity/education/dualuse/FAS_Jackson/2_A.html.

[11] Federation of American Scientists, "Mousepox Case Study," http://www.fas.org/biosecurity/education/dualuse/FAS_Jackson/4_B.html.

[12] Associated Press, " Vaccine-Evading Mousepox Virus Ignites Debate," CNN, Oct. 31, 2003.

[13] Debora MacKenzie, "US Develops Lethal New Viruses," *New Scientist,* Oct. 29, 2003, http://www.newscientist.com/article/dn4318.

[14] S.J. Robbins et. al., "The Efficacy of Cidofovir Treatment of Mice Infected with Ectromelia (Mousepox) Virus Encoding Interleukin-4," *Antiviral Research* 66, no.1 (April 2005): 1–7.

[15] Convention on the Prohibition of the Development, Production, and Stockpiling of Bacteriological (Biological) and Toxin Weapons and on their Destruction (see ch.1, note 9).

[16] Judy Miller, "Bioterrorism's Deadly Math," *City Journal,* Autumn 2008, http://www.city-journal.org/2008/18_4_bioterrorism.html.

[17] Ned Rozell, "Permafrost Preserves Clues to Deadly 1918 Flu," Alaska Science Forum, the Geophysical Institute, University of Alaska–Fairbanks, April 29, 1998, http://www.gi.alaska.edu/ScienceForum/ASF13/1386.html.

[18] Ibid.

[19] Phillip A. Sharp, "1918 Flu and Responsible Science," *Science* 310 (Oct. 7, 2005): 17.

[20] Debora MacKenzie, "Experts Fear Escape of 1918 Flu from Lab," *New Scientist,* Oct. 2004, http://www.newscientist.com/article.ns?id=dn6554.

[21] Edward Hammond, e-mail message to the Biological Weapons Prevention Project forum, Oct. 20, 2005 (with permission).

[22] Yueh-Ming Loo, Michael Gale, "Influenza: Fatal Immunity and the 1918 Virus," *Nature* 445, no.7125 (Jan. 2007): 267–68.

[23] "Lethal Secrets of 1918 Flu Virus," *BBC News,* Jan. 18, 2007, http://news.bbc.co.uk/go/pr/fr/-/2/hi/health/6271833.stm.

[24] Jens Kuhn, e-mail message to Lynn Klotz, Oct. 7, 2006.

[25] The number of deaths from the smallpox vaccine campaign is disputed. Several deaths and adverse reactions were associated with underlying heart conditions. Eight deaths coincidental to receiving the vaccine have been identified by the Department of Defense. But they claim, "These deaths were judged unrelated to vaccination, based on individual factors such as preexisting disease, incidence among unvaccinated people, and lack of physical evidence to implicate a vaccine" (Department of Defense, "Smallpox Vaccination Program Safety Summary," May 17, 2007, http://www.smallpox.army.mil/event/SPSafetySum.asp [accessed March 21, 2009]). Others dispute the Department of Defense's assertion. Vaccine expert Paul Offit "proposed a temporary halt to the vaccination program while research on a possible link between the vaccine and heart problems continues. 'Many people don't know they have underlying heart disease,' Offit said." Quoted in Anita Mannin, "Smallpox Vaccine Candidates

Narrow after Three Deaths," *USA Today,* March 31, 2003.

[26] Leonard Cole, e-mail to author, March 13, 2009, quoted with permission.

[27] Hillel W. Cohen, Robert M. Gould, and Victor Sidel, "The Pitfalls of Bioterrorism Preparedness: the Anthrax and Smallpox Experiences," *American Journal of Public Health* 94, no.10 (Oct. 2004): 1667–71, http://www.ajph.org/cgi/content/full/94/10/1667.

[28] K.H. Rubins et al., "The Host Response to Smallpox: Analysis of the Gene Expression Program in Peripheral Blood Cells in a Nonhuman Primate Model," *Proceedings of the National Academy of Science of the United States of America* 101, no.42 (Oct. 19, 2004): 15190; Marilyn Chase, "In Strictest Security, Scientists Infect Monkeys With Smallpox," *Wall Street Journal,* June 26, 2002.

[29] "Notable Quotes on Smallpox," Originally found at http://www.smallpoxbiosafety.org/quotes.html. Quoted from http://www.twnside.org.sg/title2/service160.htm.

[30] Ibid.

[31] Jeanne Guillemin, "National Security and Biodefense: Is There a Case for Full Transparency?" presentation at Pugwash Meeting 292, 21st Workshop of the Pugwash Study Group on the Implementation of the Chemical and Biological Weapons Conventions: The BWC New

Process and the Sixth Review Conference, Geneva, Switzerland, Dec. 4–5, 2004, 3.

[32] Ibid., 7.

[33] http://www.fas.org/promed/. The particular item is archive number 20050206.0404, published Feb. 6, 2005, http://www.promedmail.org/pls/otn/f?p=2400:1202:369565916594129::NO::F2400_P1202_CHECK_ DISPLAY,F2400_P1202_PUB_MAIL_ID:X,28005.

[34] Charles Piller, "Anthrax Leaks Blamed on Lax Safety Habits," *Los Angeles Times,* Aug. 20, 2004, http://www.latimes.com/news/nationworld/nation/la-na-anthrax20aug20,1,7442672.story?coll=la-home-nation.

[35] Ibid.

[36] Emily Ramshaw, "CDC Suspends A&M Research on Infectious Diseases," *Dallas Morning News,* July 1, 2007, http://www.dallasnews.com/sharedcontent/dws/dn/latestnews/stories/DN-a&m_02tex.State.Edition1.1577e7d.html.

[37] Holly Huffman, "A&M Faces Inquiry over *Brucella* Infection: University Failed to Report Researcher Exposed to Bioagent," *Eagle* (Bryan–College Station, Texas), April 18, 2007, http://209.189.226.235/stories/041807/am_20070418051.php.

[38] Emily Ramshaw, "CDC Probes A&M Bioweapons Infections," *Dallas Morning News,* June 27, 2007.

[39] Ramshaw, "CDC Suspends A&M Research on Infectious Diseases."

[40] Nick Schwellenbach, "Biodefense: A Plague of Researchers," *Bulletin of the Atomic Scientists* 61, no.3 (May/June 2005): 14–16.

[41] Michelle Fay Cortez and Jason Gale, "Baxter Sent Bird Flu Virus to European Labs by Error (Update2)," Bloomberg.com , Feb. 24 2009, http://www.bloomberg.com/apps/news?pid=20601124&sid=aiqmSoL6sVbk& refer=home (accessed March 1, 2009).

[42] Ted Sherman and Josh Margolin, "Dead Lab Mice Lost from UMDNJ Facility," *Star-Ledger* (Newark, NJ), Feb. 6, 2009.

[43] Jens Kuhn, pers. comm. with Lynn Klotz.

[44] Health and Safety Executive, "Final Report on Potential Breaches of Bio security at the Pirbright Site, 2007," Health and Safety Commission, Great Britain, 2007.

[45] Ibid., 3.

[46] Ibid., 4.

[47] "2001 United Kingdom Foot and Mouth Crisis," Wikipedia, http://en.wikipedia.org/wiki/2001_UK_foot_and_mouth_crisis; Virginia Gewin, "Agriculture Shock," *Nature* 421 (2003): 106–8.

[48] Quoted in Jim Monke, "Agroterrorism: Threats and Preparedness," Congressional Research Service, The Library of Congress, Aug. 25, 2006.

[49] Ibid.

[50] Douglas Birch, "Making, Fighting Diseases of Terror Fort Detrick: Effort Could Widen Risk, Some Experts Say," *Baltimore Sun,* June 26, 2006.

[51] Miller, "Bioterrorism's Deadly Math."

[52] Jonathan B. Tucker, Biological Threat Assessment: Is the Cure Worse Than the Disease?" *Arms Control Today,* Oct. 2004, http://www.armscontrol.org/act/2004_10/Tucker.

Chapter Seven

[1] New Oxford American Dictionary, computer application v.2.02 (emphasis added).©2005–2007.

[2] *The Public Health Security and Bioterrorism Preparedness and Response Act of 2002,* Public Law 107–188, 107th Cong. (June 12, 2002), http://www.aphis.usda.gov/programs/ag_selectagent/downloads/PL107-188.pdf (accessed March 22, 2009).

[3] *Possession, Use, and Transfer of Select Agents and Toxins; Final Rule,* 42 CFR Part 73 rev., *Federal Register Friday* (March 18, 2005), Part III, 13316–25.

[4] The URL for this report is http://www.usda.gov/oig/webdocs/33601-3-AT.pdf.

[5] Department of Health and Human Services Inspector General's office, "Summary Report on Universities' Compliance with Select Agent Regulations" (April 5, 2006), http://www.oig

.hhs.gov/oas/reports/region4/40502006.htm. Complete text of report is available at http://www.oig.hhs.gov/oas/reports/region4/40502006.pdf.

[6] *Program and Biosafety Improvement Act of 2009,* HR 1225, 111th Congress, 1st sess. (Feb. 26, 2009), http://thomas.loc.gov/cgi-bin/query/z?c111:H.R.1225:.

[7] Ibid., §201.

[8] Ibid., §203.

[9] Ibid., §203 (emphasis added).

[10] Ibid., §102.

[11] IBC Minutes Archive, The Sunshine Project, http://www.sunshineproject.org/. While the documents may be found on the Web site, the redacted material cannot be read through in the form on the Web site.

[12] The Sunshine Project, http://www.sunshine-project.org/. To access the quote, search using the term "University of North Carolina."

[13] Ibid.

[14] Ibid.

[15] Ibid.

[16] After much politicking, the grant will be awarded to Kansas State University in Manhattan Kansas (http://www.dhs.gov/xres/labs/editorial_.shtm).0762

[17] http://www.sunshine-project.org/.

[18] Committee on Research Practices and Standards to Prevent the Destructive Application of Biotechnology, "Biotechnology Research in an Age of Terrorism: Confronting the Dual Use Dilemma" (Washington, DC: National Academies Press, 2004), http://books.nap.edu/openbook.php?record_id=10827&page=1".
[19] Ibid.
[20] Ibid.
[21] National Science Advisory Board for Biosecurity, 2005–2006 Voting Members, http://oba.od.nih.gov/biosecurity/pdf/NSABB%20Voting%20Members%20Roster%202008Rev2WEB.pdf.
[22] "DRAFT Report of the NSABB Working Group on Oversight Framework Development: Proposed Strategies for Minimizing the Potential Misuse of Life Sciences Research," April 2007, http://oba.od.nih.gov/biosecurity/pdf/Framework%20for%20transmittal%200807_Sept07.pdf.
[23] Ibid. (emphasis added).
[24] Ibid.
[25] Ibid.
[26] Ibid.
[27] Ibid.
[28] Jennifer Granick, "Will Bioterror Fears Spawn Science Censorship?" *Wired,* April 25, 2007, http://www.wired.com/politics/onlinerights/commentary/circuitcourt/2007/04/circuitcourt_0425/.

[29] Ibid.

[30] John Steinbruner, Elisa D. Harris, Nancy Gallagher, Stacy M. Okutani, "Controlling Dangerous Pathogens; A Prototype Protective Oversight System," The Center for International and Security Studies at Maryland (March 2007), http://www.cissm.umd.edu/papers/files/pathogens_project_monograph.pdf.

[31] Ibid., 1–2.

[32] Ibid., 23.

[33] Ibid., 7.

[34] Ibid., 45.

[35] Granick, "Will Bioterror Fears Spawn Science Censorship?"

Chapter Eight

[1] The National Strategy for Pandemic Influenza provides over $6.1 billion for R&D over several years for stockpiling and manufacturing influenza vaccines and antivirals, with some possible additional funding from BioShield 2006. See http://www.whitehouse.gov/homeland/pandemicinfluenza.html #section1 and http://www.whitehouse.gov/homeland/pandemic-influenza-implementation.html.

[2] Lindsey Tanner, Associated Press, "Staph Fatalities May Exceed AIDS Deaths," Oct. 16, 2007.

[3] Robert G. Webster and Elizabeth Jane Walker, "Influenza," *American Scientist* 91 (March/April 2003): 122–29 (quote is from p.122).

[4] Ibid.

[5] M.T. Osterholm, "Preparing for the Next Pandemic," *New England Journal of Medicine* 352, no.18 (May 5, 2005): 1839–42.

[6] Quoted in D. Normile, "Avian Influenza: Pandemic Skeptics Warn against Crying Wolf," *Science* 310, no.5751 (Nov. 18, 2005): 1112–13.

[7] World Health Organization, "Ten Things You Need to Know about Pandemic Influenza (update of 14 Oct. 2005)," *Weekly Epidemiological Record* 80, no.49–50 (Dec. 9, 2005): 428–31.

[8] The following risk-assessment calculation first appeared in Klotz, "Casting a Wider Net for Countermeasure R&D Funding Decisions" (see ch.5, note 27).

[9] The exquisitely detailed threat assessments for bioweapons attacks now being conducted by the Department of Homeland Security also include wide ranges of estimates—because no one knows any better—for magnitudes and likelihood of bioweapons attacks. See, for example, Traci Hale, "2008 DHS Bioterrorism Risk Assessment: Planned Improvements" (PowerPoint presentations from the Committee on Methodological Improvement to the Department of Homeland Security's 2006 Bioterrorism Risk Assessment, Feb. 10, 2007); Steve Bennet, "DHS Bioterrorism Risk Assessment: Background, Requirements, and Overview,"

Aug. 28, 2006; and Richard S. Denning, "2006 DHS Bioterrorism Risk Assessment: Methodology," Aug. 28, 2006.These document are all available on the Center for Arms Control and Non-Proliferation Web site: "Documents on the Department of Homeland Security 2006 Bioterrorism Risk Assessment," http://www.armscontrolcenter.org/policy/biochem/articles/dhs_2006_bioterrorism_risk_assessment_documents/.

[10] *Project BioShield Act of 2004,* Public Law 108–276. The Strategic National Stockpile is a store of medical countermeasures against bioweapons and other threats to our national security. It is located in many places throughout the country in order to get countermeasures to victims quickly.

[11] This number is the logarithmic average of the conservative WHO estimate of five million deaths and a high estimate from Michael Osterholm of two hundred seventy million deaths.

[12] "Pandemics and Pandemic Threats since 1900," http://www.pandemicflu.gov/general/historicaloverview.html. In the United States, the 1918 pandemic flu killed 675,000 people; the 1957 pandemic flu killed about 70,000; and 1968 pandemic flu killed 33,800—not more than today's fatalities from the annual flu.

[13] This average is a minimum time period between outbreaks, because next pandemic flu could start in some future year, not 2007.

[14] This calculation first appeared in Klotz, "Casting a Wider Net for Countermeasure R&D Funding Decisions," 316.

[15] In the United States, about sixty-three deaths per hundred thousand people can be attributed to omnipresent infectious disease. For three hundred million people in the United States, this translates to one hundred seventy-seven thousand deaths from infections every year. See G.L. Armstrong, L.A. Conn, R.W. Pinner, "Trends in Infectious Disease Mortality in the United States during the 20th Century," *JAMA* 281, no.1 (Jan. 6, 1999): 61–66.

[16] Gerald L. Epstein, "Security Is More than Public Health: Commentary on 'Casting a Wider Net for Countermeasure R&D Funding Decisions,'" *Biosecurity and Bioterrorism: Biodefense Strategy, Practice, and Science* 5, no.4 (2007): p353–57.

[17] Hillel W. Cohen, Robert M. Gould, and Victor Sidel, "The Pitfalls of Bioterrorism Preparedness: The Anthrax and Smallpox Experiences," *American Journal of Public Health* 94, no.10 (Oct. 2004): 1667–71.

[18] Rhodes, "Germs, Viruses, and Secrets: The Silent Proliferation of Bio-Laboratories in the United States," Hearing of House Energy and

Commerce Committee Subcommittee on Oversight and Investigations (Oct. 4, 2007), http://energycommerce.house.gov/cmte_mtgs/110-oihrg.100407.BSL.shtml.

[19] Rhodes's testimony reported in Larry Margasak, Associated Press, "Disease-Lab Inspections Weighed for an Overhaul," *Denver Post,* Oct. 5, 2007, http://www.denverpost.com/headlines/ci_7088329.

[20] Program and Biosafety Improvement Act of 2008, 110th Congress, S. 3127, http://www.govtrack.us/congress/billtext.xpd?bill=s110-3127 ; and Select Agent Program and Biosafety Improvement Act of 2008, HR 6671, 110th Congress, http://www.govtrack.us/congress/billtext.xpd?bill=h110-6671.

[21] *Program and Biosafety Improvement Act of 2009* (see ch.7, note 6).

[22] *Project BioShield Act of 2004,* §15, p.21.

[23] Henry Cohen, "Pandemic Flu and Medical Biodefense Countermeasure Liability Legislation: P.L. 109–148, Division C (2005)," Congressional Research Service Report (April 12, 2006), http://fpc.state.gov/documents/organization/66451.pdf.

[24] The enacted version of BioShield 2006 is titled *Pandemic and All-Hazards Preparedness Act* (Public Law 109–417, 109th Congress, 2nd sess.). Sponsors and other measures related to the bill may be accessed at http://thomas.l

oc.gov/cgi-bin/bdquery/z?d109:SN03678.Apdf version of the law as passed is found at http://frwebgate.access.gpo.gov/cgi-bin/getdoc.cgi?dbname=109_cong_public_laws&docid=f:publ417.109.pdf.

[25] Public Citizen, "Willful Misconduct: How Bill Frist and the Drug Lobby Covertly Bagged a Liability Shield," *Congress Watch,* May 2006, http://www.citizen.org/documents/050406PandemicFinal_1.pdf.

[26] David Ruppe, Global Security Newswire, "Pentagon Looks to Declare Anthrax Risk Emergency, Resume Mandatory Vaccinations," Dec. 18, 2004, http://www.nti.org/d_newswire/issues/2004_12_17.html#9CF4B007.

[27] U.S. Centers for Disease Control and Prevention, "Anthrax Vaccine: What You Need to Know," http://www.cdc.gov/vaccines/pubs/vis/downloads/vis-anthrax.pdf.

[28] The NVIC is a nonprofit educational organization founded in 1982. It describes itself as dedicated to preventing vaccine injuries and deaths through public education and defending the informed consent ethic.

[29] "Vaccine Safety Advocates Oppose Pentagon's Return to Mandatory Anthrax Vaccination of U.S. Military Personnel," National Vaccine Information Center, Oct. 16, 2006. 30 David A. Geier, and Mark R. Geier, "Anthrax Vaccination: An Assessment of Potential Vaccine Safety

Concerns" (PowerPoint presentation), http://www.military-biodefensevaccines.org/.
[31] Ibid.
[32] Associated Press, "Military to Resume Mandatory Anthrax Shots: Inoculation Program to be Reinstated despite Concerns over Health Risks," Oct. 16, 2006, http://www.msnbc.msn.com/id/15294867/.
[33] Paul A. Offit "Risks of Being Risk-Averse," *Philadelphia Inquirer,* July 12, 2007.
[34] For example, see the latest report of the Pharmaceutical Research and Manufacturers of America Web site, http://www.phrma.org/files/2008%20Profile.pdf.
[35] Ibid.
[36] This and other information on the Cutter/Salk vaccine from Paul A. Offit, "Lawsuits Won't Stop Pandemics," *Wall Street Journal,* Dec. 1, 2005. Also see Paul Offit, *The Cutter Incident: How America's First Polio Vaccine Led to the Growing Vaccine Crisis* (New Haven, CT: Yale University Press, 2005).
[37] Quoted in William Tucker, "La Grippe of the Trial Lawyers," *Weekly Standard,* Oct. 25, 2004, http://www.weeklystandard.com/Content/Public/Articles/000/000/004/793dgqvs.asp.
[38] Marilyn Werber Serafini, "Vaccine Crisis Developing," *National Journal,* March 31, 2006, http://www.nti.org/d_newswire/issues/2006_3_31.html.

[39] U.S. Department of Health and Human Services, Health Resources and Services Administration, "National Vaccine Injury Compensation Program," http://www.hrsa.gov/vaccinecompensation/.
[40] Ibid.
[41] "Consensus Statement: Smallpox as a Biological Weapon," *JAMA* 281, no.22 (June 9, 1999): 2135.
[42] Dan Frosch, "Lawyer in TB Scare Is Released from Hospital," *New York Times,* July 26, 2007, http://www.nytimes.com/2007/07/26/health/27cnd-tb.html?ex=1343102400&en=48ae770306526567&ei=5088&partner=rssnyt& emc=rss.
[43] *Project BioShield Act of 2004,* §15, p.12.
[44] The term "big drug companies" refers to the forty or so major pharmaceutical company members of the Pharmaceutical Research and Manufacturers of America (http://www.phrma.org).
[45] *Pandemic and All-Hazards Preparedness Act.*
[46] Brian Gormley, Dow Jones Newswires, "Vaccine Market Draws Venture Capital Interest," Jan. 17, 2007.
[47] Chemical and Biological Defense Transformational Medical Technologies Initiative, Vol. 10, no.8, Dec. 18, 2006, http://www.dtra.mil/newsservices/fact_sheets/fs_includes/pdf/TMTI.pdf. 48 Department of Health and Human Services, Office of the Assistant Secretary for

Preparedness and Response, "HHS Public Health Emergency Medical Countermeasures Enterprise Implementation Plan for Chemical, Biological, Radiological and Nuclear Threats," *Federal Register* 72, no.77 (April 23, 2007): 20117.

[49] Ibid.

[50] Henry F. Chambers, "The Changing Epidemiology of *Staphylococcusaureus?" Emerging Infectious Diseases* 7, no.2 (Mar–Apr 2001),178–82, http://www.cdc.gov/ncidod/eid/vol7no2/pdfs/chambers.pdf ; Linda A. Johnson, Associated Press, "Pfizer, Biotech Company Team Up to Conquer Deadly Infection," Jan. 5, 2007; MRSAinfection.org, http://www.mrsainfection.org/mrsa-in-the-usa.php.

[51] *Pandemic and All-Hazards Preparedness Act,* §3678, p.91 l. 18.

[52] *Project BioShield II Act of 2005,* 109th Congress, 1st sess., §975, http://www.govtrack.us/congress/bill.xpd?bill=s109-975.Thiswas a bipartisan bill introduced by senators Joseph Lieberman (D-CT), Orrin Hatch (R-UT), and Sam Brownback (R-KS).

[53] Ibid., 350, 347.

[54] *Pandemic and All-Hazards Preparedness Act,* §3678, p.40, l. 9.

[55] *Pandemic and All-Hazards Preparedness Act,* §3678, p.41.

[56] Alan Pearson, e-mail message to Lynn Klotz, Sept. 20, 2006. Pearson was then director of

the Biological and Chemical Weapons Control Program at the Center for Arms Control and Non-Proliferation.

[57] National Security Division Directive 189, "National Policy on the Transfer of Scientific, Technical and Engineering Information" (Sept. 21, 1985), http://www.fas.org/irp/offdocs/nsdd/nsdd-189.htm.

[58] Jonathan H. Marks, "Catastrophic Opportunity: Public Health Emergencies, Health Care Infrastructure and Human Rights," chapter 18 in *Health Capital and Sustainable Socioeconomic Development,* ed. Patricia A. Cholewka and Mitra M. Motlagh (Boca Raton, FL: CRC Press with Taylor & Francis Group, 2008).

[59] Ibid p369

[60] "Protecting the Public's Health from Diseases, Disasters and Bioterrorism," Trust for America's Health, Dec. 2007, http://healthyamericans.org/reports/bioterror07/.

[61] Alan Pearson, "Fiscal Year 2009 Federal Funding for Bioweapons Prevention and Defense, Center for Arms Control and Non-Proliferation," http://www.armscontrolcenter.org/policy/biochem/articles/fy09_biodefense_funding/.

[62] Marks, "Catastrophic Opportunity," 376

[63] Unless otherwise noted, quotes from David Ozonoff, Professor Emeritus, Environmental Public Health, School of Public Health, Boston

University, are from an interview on Oct. 12, 2007.

[64] U.S. Centers for Disease Control and Prevention, "CDC Now: Protecting Health for Life," State of CDC Report 2005, http://www.scribd.com/doc/558352/CDC-Release-SOCDC2005.

Chapter Nine

[1] The Boston University laboratory is one of a handful of universityaffiliated National Centers of Excellence for Biodefense and Emerging Infectious Diseases funded by the NIH and scattered throughout the United States. There are also a few dozen university-based Regional Centers of Excellence: http://www3.niaid.nih.gov/research/resources/rce.

[2] Seattle, for example, has proposed legislation similar to Massachusetts' that goes one step further by banning BSL-4 laboratories from the city.

[3] The letter, entitled "No Place to Hide," may be downloaded from the Alternatives for Community and Environment's Web site, at http://www.ace-ej.org/BiolabWeb/Whoelse.html#Scientists,_Doctors,_Academics_.

[4] Specifically, the Conservation Law Foundation, the Boston Bar Association's Lawyers' Committee for Civil Rights Under Law, Alternatives for Community and Environment, and two law firms, McRoberts, Roberts & Rainer and Anderson & Kreiger, are working pro bono.

[5] Interview with David Ozonoff, Oct. 12, 2007.
[6] Rhodes, "High-Containment Biosafety Laboratories," 10 (see ch.1, note 5); "Germs, Viruses, and Secrets," 2 (see ch.8, note 18).
[7] Dennis Normile, "Hunt for Dengue Vaccine Heats Up as the Disease Burden Grows," *Science* 317 (Sept. 14, 2007): 1494–95.
[8] From Allen's biography on the Alston/Bannerman Fellowship Program Web site, http://www.alstonbannerman.org/2005fellows.html.
[9] "No Place to Hide," http://ace.actionmill.com/biolabweb/Biolabdocs/noplacetohide04-20-04.pdf.
[10] National Academy of Sciences, "NIH Draft Report Does Not Adequately Analyze Risks of Biocontainment Laboratory Proposed in Boston" (news release, Nov. 29, 2007).
[11] Stephen Smith, "Ruling May Stall Opening of Biolab: SJC Says Permit Wrongly Issued," *Boston Globe,* Dec. 14, 2007, http://www.boston.com/news/local/articles/2007/12/14/ruling_may_stall_opening_of_biolab/.
[12] Ibid.
[13] "NIH Outlines Next Steps to Address Safety Concerns about Boston-Area Laboratory," http://www.nih.gov/news/health/mar2008/od-06.htm. The panel is chaired by Adel Mahmoud, M.D., Ph.D., of Princeton University.
[14] Boston Public Health Commission, "Boston Public Health Commission Passes Biological

Lab Regulations" (press release, Sept. 9, 2006), http://www.bphc.org/news/press_release_content.asp ?id=363.

[15] InterAcademy Panel of the International Academy of Science, "IAP Statement on Biosecurity," Nov. 2005, http://www.interacademies.net/Object.File/Master/5/399/Biosecurity%20St..pdf.

[16] Ibid.

[17] Jeanne Guillemin, "Can Scientific Codes of Conduct Deter Bioweapons?" AlterNet, April 27, 2007, http://www.alternet.org/story/50992/.

[18] "A New Role for Scientists in the Biological Weapons Convention," *Nature Biotechnology* 23 (2005): 1213–16.

[19] Chris Schneidmiller, Global Security Newswire, "Scientific Codes of Conduct Inevitable, Experts Say," Feb. 22, 2006, http://www.nti.org/d_newswire/issues/2006/2/22/db54622f-a2c0-43fc-8eeb-5e9eeee59e62.html.

[20] Graham S Pearson and Malcolm R Dando, series eds., "Codes of Conduct for the Life Sciences: Some Insights from UK Academia" (briefing paper no.16), Strengthening the Biological Weapons Convention, Department of Peace Studies, University of Bradford, May 2005, 3, http://www.brad.ac.uk/acad/sbtwc/briefing/BP_16_2ndseries.pdf.

[21] Guillemin, "Can Scientific Codes of Conduct Deter Bioweapons?"

[22] InterAcademy Panel of the International Academy of Science, "IAP Statement on Biosecurity."

[23] K. Soeken K and D. Soeken. *A Survey of Whistleblowers: Their Stressors and Coping Strategies* (Laurel, MD: Association of Mental Health Specialties, 1987), quoted in G. Yamey, "Protecting Whistleblowers: Employers Should Respond to the Message, not Shoot the Messenger," *British Medical Journal* 320 (Jan. 8, 200): 70–71.

[24] Presentation by Henri-Philippe Sambuc, Chair, Science and Conscience Foundation, at Conscience Clause Conference, Geneva, Switzerland (Sept. 25–26, 2003), at which Lynn Klotz was a speaker. Sambuc's comments are from Klotz's notes.

[25] http://www.wikileaks.org/wiki/Wikileaks.

[26] "Wikileaks and Untraceable Document Disclosure," Jan. 3, 2007, http://www.fas.org/blog/secrecy/2007/01/wikileaks_and_untraceable_docu.html.

Chapter Ten

[1] To increase transparency of biological activities, the 1986 and 1991 review conferences of the Biological Weapons Convention adopted seven voluntary confidence-building measures that required the annual declarations and information

from the States Parties to the convention (http://wwwlaw.unimelb.edu.au/events/bwc/Confidence%20Building%20Measures.cfm)..

[2] The declaration forms for the confidence-building measures state that "States Parties agreed to provide, annually, detailed information on their biological defence research and development programmes including summaries of the objectives and costs of effort performed by contractors and in other facilities" (http://www.unog.ch/80256EDD006B8954/(httpAssets)/3CFFA8AC4E497426C12572DB00514912/$file/CBM_ Forms_Static_E.pdf [quote on p.11]).

[3] Crowley, "Disease by Design," sec. 8.3, box 9, http://www.basicint.org/pubs/Research/2001diseasebydesign1.htm (see ch.5, note 20).

[4] Ibid. Crowley is now executive director of VERTIC. VERTIC's mission is to promote effective and efficient verification as a means of ensuring confidence in the implementation of international agreements (http://www.vertic.org/aboutus.html).

[5] "Protocol to the Convention on the Prohibition of the Development, Production, and Stockpiling of Bacteriological (Biological) and Toxin Weapons and on Their Destruction," BWC/AD HOC GROUP/CRP.8 3 (April 2001), http://www.opcw.org/.

[6] The protocol also called for the establishment of an organization to administer the protocol. A

subset of States Parities called the Executive Council would make official decisions, such as launching facility investigations.

[7] Douglas J MacEachin, "Routine and Challenge: Two Pillars of Verification," *CBW Conventions Bulletin* 39 (March 1998): 4.

[8] The Chemical Manufacturers Association has changed its name to the American Chemistry Council. It is the Washington-based lobbying organization for the chemical industry.

[9] "Convention on the Prohibition of the Development, Production, Stockpiling and Use of Chemical Weapons and on their Destruction."

[10] The elements of Managed Access may be found in "Access and Measures to Guard against Abuse during the Conduct of Investigations," sec. G. In contrast to the Chemical Weapons Convention, the words "Managed Access" are not used in the protocol.

[11] John Gilbert, e-mail message to Lynn Klotz, April 2008.

[12] Ibid.

[13] Greg Seigle, Global Security Newswire, "BWC: U.S. Pressuring Several Countries to Comply With Treaty," Jan. 11, 2002, http://www.nti.org/d_newswire/issues/newswires/2002_1_11.html#8.

[14] MacEachin, "Routine and Challenge," 2. As the protocol was being developed, several different terms, including "non-challenge visits," were

used to describe investigations and visits before settling on the more politically acceptable terms "facility investigations" and "transparency visits."

[15] MacEachin cites Alan Zelicoff (spelled "Zelikoff" in the original article), "Be Realistic about Biological Weapons," *Washington Post,* Jan. 8, 1998. MacEachin, "Routine and Challenge," 3.

[16] Ibid.

[17] Lynn C. Klotz et al., "Implementing the Biological Weapons Protocol in the United States: What It Means to the Biopharmaceutical Industry," *BioPharm,* Aug. 2000, 46–48. Lynn C. Klotz and B.H. Rosenberg, "A Means for Protecting US Industry Within an Effective Compliance Regime for the Biological Weapons Convention" Federation of American Scientists Working Group on Biological and Toxin Weapons Verification working paper, 1999).

[18] The Federation of American Scientists and the Pharmaceutical Research and Manufacturers of America conducted several discussions on the protocol. The goal of the FAS was to understand PhRMA's concerns over the protocol and to persuade them to support it. The discussions resulted in a joint publication, Klotz et al., "Implementing the Biological Weapons Protocol in the United States."

[19] Convention on the Prohibition of the Development, Production, and Stockpiling of Bacterio-

logical (Biological) and Toxin Weapons and on their Destruction, April 10, 1972, article X, para. 2.

[20] Prior to 2001, terrorists were often referred to as nonstate actors, a term that somehow invokes less fear, and that was the language used in the earlier protocol discussion.

[21] The Australia Group, http://www.australiagroup.net/en/index.html.

[22] One of the ways in which the chairman's text of the protocol (see note 10, above) was weakened to get U.S. support was to exempt pharmaceutical facilities from visits.

[23] Richard Weitz, "International Unit is Underappreciated Tool in Fight Against Bio-Terror, *World Politics Review,* Jan. 11, 2008, http://www.worldpoliticsreview.com/article.aspx?id=1494. Weitz is the director of the Center for Political-Military Analysis Senior Fellow Hudson Institute in Washington, D.C.

[24] Robert H. Sprinkle, "The Biosecurity Trust," *BioScience* 53, no.3 (March 2003): 270–76.

[25] Ibid., 275.

[26] Matthew Meselson and Julian Robinson, "A Draft Convention to Prohibit Biological and Chemical Weapons under International Criminal Law," Harvard Sussex Program on CBW Armament and Arms Limitation, Nov. 1, 2001. The actual treaty title is "Draft Convention on the Prevention and Punishment of the Crime of

Developing, Producing, Acquiring, Stockpiling, Retaining, Transferring, or Using Biological or Chemical Weapons," and it may be found at www.fas.harvard.edu~hsp/crim01.pdf./

[27] "Convention for the Suppression of Unlawful Seizure of Aircraft," Oct. 14, 1971, http://cns.miis.edu/pubs/inven/pdfs/airseiz.pdf.

[28] Meselson and Robinson, "A Draft Convention to Prohibit Biological and Chemical Weapons under International Criminal Law," 1.

[29] Ibid.

[30] Michael Barletta, Amy Sands, Jonathan B. Tucker, "Keeping Track of Anthrax: The Case for a Biosecurity Convention," *Bulletin of the Atomic Scientists* 58, no.3 (May/June 2002): 57–62, http://thebulletin.metapress.com/content/y6216t53k2t0x2lt/.

[31] John Steinbruner, Elisa D. Harris, Nancy Gallagher, Stacy M. Okutani, "Controlling Dangerous Pathogens; A Prototype Protective Oversight System," The Center for International and Security Studies at Maryland, March 2007.

[32] "Compliance Review: An Important National Implementation Measure" (statement prepared for the 2007 Meeting of Experts of the BWC by the Center for Arms Control and Non-Proliferation). Contact Alan Pearson for a copy of his remarks: ampearson2[at]gmail.com.

[33] "The Sixth Review Conference of the States Parties to the Convention on the Prohibition of

the Development, Production and Stockpiling of Bacteriological (Biological) and Toxin Weapons and on their Destruction, Geneva, 20 November–8 December 2006" (working Paper no.3, Assessment of National Implementation of the Biological and Toxin Weapons Convention (BTWC), Oct. 20, 2006).

[34] This famous statement was made by Eldridge Cleaver in a 1968 speech in San Francisco. Eldridge Cleaver, *Post-Prison Writings and Speeches,* ed. R. Scheer (New York: Random House, 1969). This quote was found in Andrews, Robert; Biggs, Mary; and Seidel, Michael, et al., ed., *The Columbia World of Quotations* (New York: Columbia University Press, 1996), http://www.bartleby.com/66/14/12614.html.

[35] Our definition was expanded from the Proposed Senate Bill S.1873, reported to the Senate by Mr. Enzi on October 24, 2005. There is a movement to narrow the definition of biosecurity to only biodefense. "Biosafety" would then be the term used to describe lab safety, accidents, accidental releases of pathogens, etc. "Dual-use" would describe science and technologies that can be exploited for good or hostile purposes. Natural infectious diseases would not be a concern of "biosecurity," under these definitions. One reason for narrowing the definition of biosecurity to biodefense only is to

avoid confusion. There is also concern that public health will get swept up in biosecurity regulations because the broad definition "could tie public health too closely to national security agendas and may threaten the freedom of scientific research"(Anna Zmorzynska and Iris Hunger, "Restricting the Role of Biosecurity," *Bulletin of the Atomic Scientists* Dec. 19, 2008, http://thebulletin.org/web-edition/features/restricting-the-role-of-biosecurity). Protecting scientific freedom and safeguarding public health issues from being obscured by the national security agenda are important goals. However, we believe that to achieve those very goals—and others we have discussed—all infectious disease- related public health must be addressed together. We argue for the broadest defi nition of biosecurity that encompasses biodefense, biosafety, dual use, and natural infectious diseases.

Epilogue

[1] "Reducing Biological Risks to Security: International Policy Recommendations for the Obama Administration," Jan. 2008 [2009], http://www.armscontrolcenter.org/assets/pdfs/biothreats_initiatives.pdf (accessed May 26, 2009).

[2] Christine Todd Whitman, "Your Inbox, Mr. President/Six Leading Voices Tell Nature What the New US President Needs to do to Move beyond

the Bush Legacy," *Nature* 457 (Jan. 15, 2009): 258–61.

[3] Tara O'Toole and Thomas Inglesby, "Biosecurity Memos to the Obama Administration," *Biosecurity and Bioterrorism: Biodefense Strategy, Practice, and Science* 7, no.1 (March 2009): 25–28.

[4] "President Obama Announces Members of Science and Technology Advisory Council," White House, Office of the Press Secretary, April 27, 2009, http://www.whitehouse.gov/the_press_office/President-Obama-Announces-Members-of-Science-and-Technology-Advisory-Council/(accessed May 26, 2009).

[5] Christopher Chyba, "Biotechnology and the Challenge to Arms Control, "*Arms Control Today,* October 2006, http://www.armscontrol.org/act/2006_10/BioTechFeature (accessed May 26, 2009).

[6] Before the Nuclear Threat Initiative, Hamburg was assistant secretary for planning and evaluation in the Department of Health and Human Services during the Clinton administration. She also spent six years as New York City health commissioner and was the assistant director of the National Institute of Allergy and Infectious Diseases at the National Institutes of Health.

[7] Klotz et al., "Implementing the Biological Weapons Protocol in the United States" (see ch.10, note 17).

A

B

C

D

E

F

J

K

L

T

U

V

W

Y

Books For ALL Kinds of Readers

At ReadHowYouWant we understand that one size does not fit all types of readers. Our innovative, patent pending technology allows us to design new formats to make reading easier and more enjoyable for you. This helps improve your speed of reading and your comprehension. Our EasyRead printed books have been optimized to improve word recognition, ease eye tracking by adjusting word and line spacing as well as minimizing hyphenation. Our EasyRead SuperLarge editions have been developed to make reading easier and more accessible for vision-impaired readers. We offer Braille and DAISY formats of our books and all popular E-Book formats.

Printed in Dunstable, United Kingdom

91338351R00248